iPhone® 4S For Seniors

FOR
DUMMIES®

D1506910

by Nancy Muir

WILEY

John Wiley & Sons, Inc.

iPhone® 4S For Seniors For Dummies®

Published by
John Wiley & Sons, Inc.
111 River Street
Hoboken, NJ 07030-5774

www.wiley.com

For general information on our other products and services, please contact our Customer Care Department within the U.S. at 877-762-2974, outside the U.S. at 317-572-3993, or fax 317-572-4002.

For technical support, please visit www.wiley.com/techsupport.

Wiley publishes in a variety of print and electronic formats and by print-on-demand. Some material included with standard print versions of this book may not be included in e-books or in print-on-demand. If this book refers to media such as a CD or DVD that is not included in the version you purchased, you may download this material at http://booksupport.wiley.com. For more information about Wiley products, visit www.wiley.com.

Library of Congress Control Number: 2011944109

ISBN 978-1-118-20961-5 (pbk); ISBN 978-1-118-22870-8 (ebk); ISBN 978-1-118-23137-1 (ebk); ISBN 978-1-118-26605-2 (ebk)

Manufactured in the United States of America

10 9 8 7 6 5 4 3 2

WILEY

About the Author

Nancy Muir is the author of over 60 books on technology and business topics. In addition to her writing work, Nancy runs a website on technology for seniors called `TechSmartSenior.com` and a companion website for her iPad books in the *For Dummies* series, `iPadMadeClear.com`. She writes a regular column on computers and the Internet on Retirenet. com. Prior to her writing career Nancy was a manager at several publishing companies and a training manager at Symantec.

Dedication

To Steve Jobs, who passed away shortly after iPhone 4S was announced. He introduced elegant technology in such a way that consumers like you and I could see the fun in it all.

Author's Acknowledgments

I was lucky enough to have Blair Pottenger, the absolute best editor in the world, assigned to lead the team on this book. Blair, I couldn't have gotten through this rush schedule without you; you went way above and beyond on this one! Thanks also to Dennis Cohen for his able work as technical editor, and to Heidi Unger, the book's copy editor. Last but never least, thanks to Kyle Looper, Acquisitions Editor, for hiring me to write this book.

Publisher's Acknowledgments

We're proud of this book; please send us your comments at http://dummies.custhelp.com. For other comments, please contact our Customer Care Department within the U.S. at 877-762-2974, outside the U.S. at 317-572-3993, or fax 317-572-4002.

Some of the people who helped bring this book to market include the following:

Acquisitions and Editorial

Project Editor: Blair J. Pottenger

Acquisitions Editor: Kyle Looper

Copy Editor: Heidi Unger

Technical Editor: Dennis Cohen

Editorial Manager: Kevin Kirschner

Editorial Assistant: Amanda Graham

Sr. Editorial Assistant: Cherie Case

Cover Photo: © iStockphoto.com/Nicolas Hansen

Cartoons: Rich Tennant (www.the5thwave.com)

Composition Services

Project Coordinator: Katie Crocker

Layout and Graphics: Carl Byers, Joyce Haughey, Christin Swinford

Proofreaders: Melissa D. Buddendeck, Lauren Mandelbaum

Indexer: BIM Indexing & Proofreading Services

Publishing and Editorial for Technology Dummies

Richard Swadley, Vice President and Executive Group Publisher

Andy Cummings, Vice President and Publisher

Mary Bednarek, Executive Acquisitions Director

Mary C. Corder, Editorial Director

Publishing for Consumer Dummies

Kathleen Nebenhaus, Vice President and Executive Publisher

Composition Services

Debbie Stailey, Director of Composition Services

Table of Contents

*I*f you bought this book (or are even thinking about buying it), you've probably already made the decision to buy an iPhone. The iPhone is set up to be simple to use, but still, you can spend hours exploring the preinstalled apps, finding how to change settings, and syncing the device to your computer or through iCloud. I've invested those hours so that you don't have to — and I've added advice and tips for getting the most out of your iPhone.

This book helps you get going with the iPhone quickly and painlessly so that you can move directly to the fun part.

About This Book

This book is specifically written for mature people like you, folks who may be relatively new to using a smartphone and want to discover the basics of buying an iPhone, working with its preinstalled apps, and getting on the Internet. In writing this book, I've tried to consider the types of activities that might interest someone who is 50 years old or older and picking up an iPhone for the first time.

Conventions Used in This Book

This book uses certain conventions to help you find your way around, including

➡ Text you type in a text box is in **bold**. Figure references, such as "see **Figure 1-1**," are also in bold, to help you find them.

➡ Whenever I mention a website address, or *URL*, I put it in a different font, like this.

➡ Figure callouts draw your attention to actions you need to perform. In some cases, points of interest in a figure might be indicated. The text tells you what to look for; the callout line makes it easy to find.

Tip icons point out insights or helpful suggestions related to tasks in the step lists.

New icons highlight what features of iPhone or iOS 5 are new and exciting, in case you're moving up from earlier versions.

Foolish Assumptions

This book is organized by sets of tasks. These tasks start from the beginning, assuming that you've never laid your hands on an iPhone, and guide you through basic steps provided in nontechnical language.

This book covers going online using either a Wi-Fi or 3G connection. I'm also assuming that you'll want to download and use the iBooks e-reader app, so I tell you how to download it in Chapter 13 and cover its features in Chapter 14.

Why You Need This Book

The iPhone is cool and perfect for many seniors because it provides a simple, intuitive interface for making calls, but also for activities such as checking e-mail and playing music. But why should you stumble around, trying to figure out its features? Following the simple, step-by-step approach used in this book, you can get up to speed with the iPhone right away and overcome any technophobia you might have.

How This Book Is Organized

This book is conveniently divided into several handy parts to help you find what you need:

➡ **Part I: Making the iPhone Work for You:** If you're about to buy your iPhone or are ready to get started with the basics of using it, this part is for you. These chapters highlight the newest features in iPhone and iOS 5 and help you explore the different specifications, styles, and price ranges for all iPhone models. You find out how to set up your iPhone out of the box, including

  • Opening an iCloud account to register and push content to all your Apple devices automatically.

  • Opening an iTunes account to buy entertainment content and additional apps.

These chapters also provide information for exploring the iPhone Home screen when you first turn the phone on, and useful accessibility features to help out if you have hearing or vision challenges.

➡ **Part II: Start Using Your iPhone:** In this part, you learn the basics of making and receiving calls — this is a phone, after all! You discover how to set up and manage your contacts, how to get the most out of some useful utility apps, and how to take advantage of iPhone's accessibility features.

In this part, you're also introduced to Siri, the iPhone 4S's hot new feature that allows you to talk to your phone and have it provide information and perform tasks for you. You also explore the exciting FaceTime feature, used for making video calls to other people who use the iPhone 4 or 4S, iPad 2, a Mac, or the iPod touch.

 Finally, you discover how integrated features for using Twitter and iMessage (the latter is accessed via the Messages instant-messaging app) help you to connect with others.

➡ **Part III: Taking the Leap Online:** Here, you find out how to connect to the Internet and use the built-in Safari browser. You putter with the preinstalled Mail app and set up your iPhone to access e-mail from your existing e-mail accounts. In this part, you also get to shop online at the iTunes Store for multimedia content, such as movies and music, and shop the App Store for additional iPhone apps.

➡ **Part IV: Having Fun and Consuming Media:** The iPhone has been touted by some as a great device for consuming media such as music, podcasts, and movies. Included with every iPhone are a Music app for playing music and the Videos and YouTube apps for watching video content. In addition, in this part I explain how to use iBooks, the free e-reader app from Apple. You also explore playing games on your iPhone, which — trust me — is a lot of fun, and I help you experiment with the Maps app to find your favorite restaurant or bookstore with ease. You also discover the wonderful possibilities for using still and video cameras on iPhone.

In this part, you also explore the new Newsstand app for subscribing to and reading magazines.

⟹ **Part V: Managing Your Life and Your iPhone:** For the organizational part of your brain, the iPhone makes available Calendar, Notification Center, Reminders, and Notes apps, all of which are covered in this part. I also offer advice about keeping your iPhone safe and troubleshooting common problems that you might encounter, including using the Find My iPhone feature to deal with a lost or stolen iPhone. You can also use the new iCloud service to back up your content or restore your iPhone.

The new Reminders app and Notification Center feature in iOS 5 are great for keeping you on schedule. Reminders is a great to-do list feature that allows you to enter tasks and details about them, and can also display tasks from your online calendars. Notification Center lists all your alerts and reminders in one place.

Where to Go from Here

You can work through this book from beginning to end or simply open a chapter to solve a problem or acquire a specific new skill whenever you need it. The steps in every task quickly get you to where you want to go, without a lot of technical explanation.

Note: At the time I wrote this book, all the information it contained was accurate for the iPhone 3GS, iPhone 4, and iPhone 4S, version 5 of the iOS (operating system) used by the iPhone, and version 10.5 of iTunes. Apple is likely to introduce new iPhone models and new versions of iOS and iTunes between book editions. If you've bought a new iPhone and its hardware, user interface, or the version of iTunes on your computer looks a little different, be sure to check out what Apple has to say at `www.apple.com/iphone`. You'll no doubt find updates on the company's latest releases. When a change is very substantial, we may add an update or bonus information that you can download at this book's companion website, `www.dummies.com/go/iphoneforseniors`.

Part I

Making the iPhone Work for You

The 5th Wave   By Rich Tennant

"Okay antidote, antidote, what would an antidote app look like? You know, I still haven't got this Home screen the way I'd like it."

Buying Your iPhone

*Y*ou've read about it. You've seen on the news the lines at Apple Stores on the day a new version of the iPhone is released. You're so intrigued that you've decided to get your own iPhone to have a smartphone that offers more than making and receiving calls. iPhone also offers lots of fun apps; allows you to explore the online world; read e-books, magazines, and periodicals; organize your photos, and more.

Trust me: You've made a good decision, because the iPhone redefines the mobile phone experience in an exciting way. It's also an absolutely perfect fit for many seniors.

In this chapter, you learn about the advantages of iPhone, as well as where to buy this little gem and associated data plans from providers. After you have one in your hands, I help you explore what's in the box and get an overview of the little buttons and slots you'll encounter — luckily, the iPhone has very few of them.

Get ready to . . .

Discover What's New in iPhone 4S and iOS 5

Apple's iPhone gets its features from a combination of hardware and its software operating system (called *iOS*; the term is short for iPhone operating system, in case you need to know that to impress your friends). The most current operating system is iOS 5. It's helpful to understand which new features the iPhone 4S device and iOS 5 bring to the table (all of which are covered in more detail in this book). New features in iPhone 4S include

➠ **Siri personal assistant:** This voice recognition system armed with artificial intelligence can provide intelligent responses to your spoken questions or act on your request, such as reminding you about an appointment.

➠ **A dual-core A5 chip:** This chip gives your iPhone much faster performance.

➠ **All-new camera:** An improved, 8-megapixel camera with an improved light sensor.

➠ **The ability to shoot high-definition video:** This feature includes a stabilization feature that helps keep those shakes out of your video.

 Throughout this book, I highlight features that are relevant only in using the iPhone 4S, so you can use this book no matter which version of the iPhone you own as long as you have iOS 5 installed.

In addition, your iPhone may have iOS 4 *or* 5 installed, depending on which phone model you bought. Any iPhone device more recent than the iPhone 3G can make use of iOS 5 if you update the operating system (discussed in detail in Chapter 2); this book is based on version 5 of iOS. This update to the operating system adds many new features, including

➠ **Integration with iCloud,** including the ability to back up and restore your iPhone. *iCloud* is a new service from Apple that allows you to save and retrieve files from an online account, sync content with other Apple devices, and update your iPhone operating system without having to sync your device to your computer.

➡ **Newsstand,** an app that allows you to subscribe to and read online versions of many popular magazines and newspapers that are *pushed* to your iPhone so you have the latest editions without having to do a thing once you've bought a subscription.

➡ **Reminders,** a great place to centralize all your upcoming events, set reminders, and organize your commitments by date or in a list format. You can also have iPhone remind you to take actions when you leave or arrive at a location (leaving the grocery store, remember to call your spouse to ask if there's anything else for you to get!).

➡ Notifications delivered in the **Notification Center,** where you can control how iPhone lets you know about FaceTime alerts, new messages or reminders, events in your Calendar, and items such as badges, sounds, and banners.

➡ **iMessage,** a new, integrated instant-messaging app for sending text messages to people using other Apple devices in real time. (Now you send it; now they see it.)

➡ Additional **touchscreen gestures** that provide short-cuts for getting things done, such as dragging e-mail addresses to address fields in Mail and multitasking.

➡ **Accessibility features** such as LED flash and vibra-tion settings that help to alert those with hearing or vision challenges to incoming calls or messages.

➡ **Integration with Twitter** from several apps — including Photos, Maps, and the Safari web browser.

➡ **E-mail tools** that allow you to apply bold, italic, underlining, and indentation settings to your e-mail messages, as well as improved searching of messages.

➡ **PC Free,** which is all about liberating your device from your computer so you can control many actions — such as updating to the latest version of iOS — directly from iPhone without syncing to your computer and iTunes.

➡ New **Game Center** features such as posting profile pictures, playing turn-based games, and helping you compare your scores with your friends'.

Choose the Right iPhone for You

iPhones don't come in different sizes. In fact, if you pick up an iPhone 4 and 4S (see **Figure 1-1**), you're not likely to be able to tell one model from another, except that some are black and some are white. (iPhone 3GS is a little longer and has slightly rounded edges, but the later phones are identical.) Their differences are primarily under the hood.

iPhone 4S models have two variations:

➡ Black or white

➡ Amount of built-in memory ranging from 16GB to 64GB.

Your options in the first bullet point are pretty black and white, but if you're confused about the other one, read on as I explain these variations in more detail in the following sections.

Table 1-1 gives you a quick comparison of iPhone 3GS, 4, and 4S. All costs are as of the time this book was written.

Figure 1-1

Table 1-1	iPhone Model Comparison				
Model	*Memory*	*Cost (with a 2-Year Contract)*	*Support for FaceTime*	*Siri*	*Carriers*
3GS	8GB	Free	No	No	AT&T
4	8GB	$99	Yes	No	AT&T, Verizon, Sprint
4S	16–64 GB	$199–$399	Yes	Yes	AT&T, Verizon, Sprint

Decide How Much Memory Is Enough

Memory is a measure of how much information — for example, movies, photos, and software applications (apps) — you can store on a computing device. Memory can also affect your iPhone's performance when handling tasks such as streaming favorite TV shows from the World Wide Web or downloading music.

Streaming refers to watching video content from the web (or from other devices) rather than playing a file stored on your computing device. You can enjoy a lot of material online without ever downloading its full content to your hard drive — and given that every iPhone model has a relatively small amount of memory, that's not a bad idea. See Chapters 15 and 17 for more about getting your music and movies online.

Your memory options with an iPhone are 8G, or 16, 32, or 64 gigabytes (GB) (new with iPhone 4S). You must choose the right amount of memory because you can't open the unit and add memory, as you usually can with a desktop computer. However, Apple has thoughtfully provided iCloud, a new service you can use to back up content to the Internet (you can read more about that in Chapter 3).

So how much memory is enough for your iPhone? Here's a rule of thumb: If you like lots of media, such as movies or TV shows, you might need 64GB. For most people who manage a reasonable number of photos, download some music, and watch heavy-duty media such as movies online, 32GB is probably sufficient. If you simply want to check e-mail, browse the web, and write short notes to yourself, 16GB *might* be enough.

 Do you have a clue how big a gigabyte (GB) is? Consider this: Just about any computer you buy today comes with a minimum of 250GB of storage. Computers have to tackle larger tasks than iPhones do, so that number makes sense. The iPhone, which uses a technology called *flash* for memory storage, is meant (to a great extent) to help you experience online media and e-mail; it doesn't have to store much and in fact pulls lots of content from online. In the world of memory, 16GB for any kind of storage is puny if you keep lots of content and graphics on the device.

What's the price for larger memory? For the iPhone 4S, a 16GB unit costs $199 with a two-year contract; 32GB jumps the price to $299; and 64GB adds another $100, setting you back a whopping $399.

Understand What You Need to Use Your iPhone

Before you head off to buy your iPhone, you should know what other connections and accounts you'll need to work with it optimally.

At a bare minimum, to make standard cellular phone calls you need to have a service plan with a cellular carrier such as AT&T, as well as a data plan that supports iPhone. The data plan allows you to exchange data over the Internet.

You also need to be able to update the iPhone operating system and share media such as music among Apple devices. Though these things can be done without a phone carrier service plan you have to plug your phone into your computer to update the iOS or use a local Wi-Fi network to go online and make calls using an Internet service such as Skype. Given the cost and hi-tech nature of the iPhone, having to jury-rig these basic functions doesn't make much sense so trust me, get an account and data plan.

You can open an iCloud account to store and share content online, or you can use a computer to download photos, music, or applications from non-Apple online sources such as stores or sharing sites like your local library and transfer them to your iPhone through a process called *syncing*. You can also use a computer or iCloud to register your iPhone the first time you start it, although you can have the folks at the Apple Store, AT&T, Sprint, or Verizon handle registration for you if you have one nearby.

Apple has set up its iTunes software and the iCloud service to give you two ways to manage content for your iPhone — including movies, music, or photos you've downloaded — and specify how to sync your calendar and contact information. Chapter 3 covers those settings in more detail.

Know Where to Buy Your iPhone

You can't buy iPhone from every major retail store such as Sears. You can buy an iPhone at the Apple Store and from mobile phone providers AT&T, Sprint, and Verizon. You can also find an iPhone at major retailers such as Best Buy and Walmart, through whom you have to buy a two-year service contract for the phone carrier of your choice. You can also find iPhone at several online retailers such as Amazon.com and Newegg.com.

 Apple offers unlocked iPhones that can be used with any of the three iPhone cellular service providers, but these phones without accompanying phone plans can be pretty pricey.

Explore What's in the Box

When you fork over your hard-earned money for your iPhone, you'll be left holding one box about the size of a deck of tarot cards. Here's a rundown of what you'll find when you take off the shrink-wrap and open the box:

➡ **iPhone:** Your iPhone is covered in a thick, plastic sleeve-thingie that you can take off and toss (unless you think there's a chance you'll return it, in which case you might want to keep all packaging for 14 days — Apple's standard return period).

➡ **Apple earphones with remote and mic:** Plug these earphones into your iPhone for a free headset experience.

➡ **Documentation (and I use the term loosely):** Notice, under the iPhone itself, a small, white envelope about the size of a half-dozen index cards. Open it and you'll find:

- *A tiny pamphlet:* This pamphlet, named *Important Product Information Guide*, is essentially small print (that you mostly don't need to read) from folks like the FCC.

- *A label sheet:* This sheet has two white Apple logos on it. (I'm not sure what they're for, but my husband and I use one sticker to differentiate my iPhone from his.)

- *A small foldout card:* This card provides panels containing photos of the major features of iPhone 4S and information about where to find out more. (Prior to 4S you got only a single card with a photo of the phone and callouts to major features; 4S documentation expanded exponentially . . . which isn't saying much!).

➡ **Dock Connector to USB Cable:** Use this cord (see **Figure 1-2**) to connect the iPhone to your computer, or use it with the last item in the box, the USB Power Adapter.

Dock Connector to USB Cable   10W USB Power Adapter

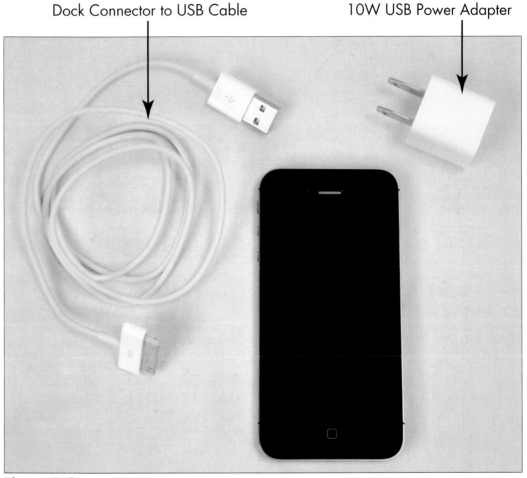

Figure 1-2

➡ **USB Power Adapter:** The power adapter (refer to **Figure 1-2**) attaches to the Dock connector cord so that you can plug it into the wall and charge the battery.

That's it. That's all there is in the box. It's kind of a study in Zen-like simplicity.

 Try searching for iPhone accessories online. You'll find iPhone cases ranging from leather to silicone; car chargers; and screen guards to protect your phone's screen.

Take a First Look at the Gadget

The little card contained in the documentation (see the preceding section) gives you a picture of the iPhone with callouts to the buttons you'll find on it. In this section, I give you a bit more information about those buttons and other physical features of the iPhone. **Figure 1-3** shows you where each of these items is located.

Here's the rundown on what the various hardware features are and what they do:

➡ **(The all-important) Home button:** On the iPhone, you can go back to the Home screen to find just about anything. The Home screen displays all your installed and preinstalled apps and gives you access to your iPhone settings. No matter where you are or what you're doing, push Home and you're back at home base. You can also double-tap the Home button to pull up a scrolling list of apps so you can quickly move from one to another.

➡ **On/Off button:** You can use this button (whose functionality I cover in more detail in Chapter 2) to power up your iPhone, put it in Sleep mode, wake it up, or power it down.

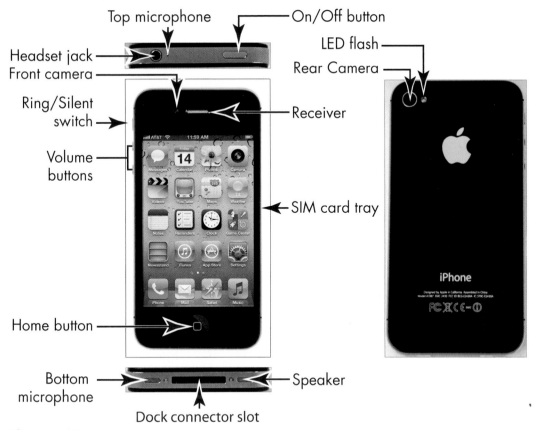

Top microphone | On/Off button
Headset jack | LED flash
Front camera | Rear Camera
Ring/Silent switch | Receiver
Volume buttons | SIM card tray
Home button
Bottom microphone | Speaker
Dock connector slot

Figure 1-3

⟶ **Dock connector slot:** Plug in the Dock Connector to USB Cable to charge your battery or sync your iPhone with your computer (which you find out more about in Chapter 3).

⟶ **Cameras:** The iPhone 4 and 4S offer front- and rear-facing cameras that you can use to shoot photos or video. The rear one is on the top-right corner, and you need to be careful not to put your thumb over it when taking shots (I have several very nice photos of my thumb already).

⟶ **Ring/Silent switch:** Slide this little switch to mute or unmute the sound on your iPhone.

⟹ **(A tiny, mighty) Speaker:** One nice surprise when I first got my iPhone was hearing what a nice little sound system it has and how much sound can come from this tiny speaker. The speaker is located on the bottom edge of the phone, below the Home button.

⟹ **Volume buttons:** Tap the volume up button for more volume and the volume down button for less. With iOS 5, you can use the volume up button as a camera shutter button when the camera is activated.

⟹ **Headset jack and microphones:** If you want to listen to your music in private, you can plug in the iPhone headset (which gives you bidirectional sound) or 3.5mm minijack headphones. Two microphones (one each on the top and bottom of the phone) makes it possible to speak into your iPhone to do things such as make phone calls using the Internet, video calling services, or other apps that accept audio input such as Siri built-in assistant.

⟹ **Receiver:** The mesh on the top front of the device that you hear through.

⟹ **SIM card tray:** The little slot for placing your phone carrier's SIM card which allows your phone to connect to the carrier and stores data such as contacts in memory.

⟹ **LED flash:** The flash device for the built-in cameras.

Looking Over the Home Screen

1 won't kid you: You have a slight learning curve ahead of you because iPhone is different from other mobile phones you may have used (although, if you own another smartphone, you've got a head start). For example, your previous phone might not have had a Multi-Touch screen and onscreen keyboard.

The good news is that getting anything done on the iPhone is simple, once you know the ropes. In fact, using your fingers instead of a tiny physical keyboard to do things onscreen is a very intuitive way to communicate with your computing device, which is just what iPhone is.

In this chapter, you turn on your iPhone and register it and then take your first look at the Home screen. You also practice using the onscreen keyboard, see how to interact with the touchscreen in various ways, get pointers on working with cameras, and get an overview of built-in applications.

Get ready to . . .

Have a soft cloth handy, like the one you might use to clean your eyeglasses. You're about to deposit a ton of fingerprints on your iPhone — one downside of a touchscreen device.

See What You Need to Use iPhone

You need to be able, at minimum, to connect to the Internet to take advantage of most iPhone features, which you can do using a Wi-Fi network or 3G from your cellular provider. You might want to have a computer so that you can download photos, videos, music, or applications and transfer them to your iPhone through a process called *syncing* (see Chapter 3 for more about syncing). With iOS 5, a new Apple service called iCloud has arrived, which syncs content from all your Apple iOS devices so anything you buy on your iPad, for example, will automatically be pushed to your iPhone.

Your phone will probably arrive registered and activated or, if you buy it in a store, the person helping you can handle that procedure.

For an iPhone 4S, Apple's *iPhone User Guide* recommends that you have

➥ A Mac or PC with a USB 2.0 port and one of these operating systems:

- Mac OS X Lion version 10.7.2 or later

- Windows 7 or Windows Vista

➥ iTunes 10.5 or later, available at www.itunes.com/download

➥ An iTunes Store account

➥ Internet access

Apple has set up its iTunes software to help you manage content for your iPhone — which includes the movies, music, or photos you've downloaded — and specify from where to transfer your calendar and contact information. Chapter 3 covers these settings in more detail.

Turn On iPhone

1. The first time you turn on your iPhone, it will probably have been activated and registered by your phone carrier or Apple (if you buy it from Apple).

2. Press and hold the On/Off button on the top of your iPhone until the Apple logo appears. In another moment, a screen appears asking if you'd like to register via iCloud or use iTunes. Tap the iTunes option and proceed. (If, instead, you'd like to use iCloud, see the next task.)

3. Plug the Dock Connector to USB Cable that comes with your device into your iPhone.

4. Plug the other end of the cable into a USB port on a computer. Both your computer and the iPhone think for a few moments while they exchange data.

5. Sign in with your Apple ID in the dialog that appears on your computer screen, and then follow the simple onscreen instructions in subsequent screens to register your iPhone and choose the content to automatically download when you connect the iPhone to your computer. (You can change these settings later; these steps are covered in Chapter 3.) When you're done, your iPhone Home screen appears, and you're in business.

6. Unplug the Dock Connector to USB Cable.

 If you buy your iPhone at an Apple Store, an employee will register it for you, and you can skip this whole process.

 You can choose to have certain items transferred to your iPhone from your computer when you sync: music, videos, downloaded applications, contacts, audiobooks, calendars, e-books, podcasts, and browser bookmarks. You can also transfer to your computer any content you download directly to your iPhone using the iTunes and App Store apps or non-Apple stores. See Chapters 12 and 13 for more about these features.

Register PC-Free Using iCloud

In Step 2 of the previous task, you can choose to register your device via iCloud. To use this PC-free process, you need to be within range of a Wi-Fi hotspot or use your iPhone's 3G connection so you can connect directly to the Internet.

When you make this choice, rather than signing into iTunes you provide an Apple ID. If you don't have an Apple ID, you're offered the option of creating one right then and there. That ID is associated with your iCloud account, and you can use it for various iCloud-supported activities.

You'll also be asked to respond to various questions, such as your preferred language and country. When you finish answering these, your iPhone is registered without ever plugging it into a computer.

Meet the Multi-Touch Screen

When the iPhone Home screen appears (see **Figure 2-1**), you see a pretty background and two sets of icons. One set appears in the Dock, along the bottom of the screen. The *Dock* contains the Phone, Mail, Safari, and Music app buttons by default, though you can swap out one app for another. The Dock appears on every Home screen. You can add new apps to populate as many as 11 additional Home screens. Other icons appear above the Dock and are closer to the top of the screen. (I cover all these icons in the "Take Inventory of Preinstalled Applications" task, later in this chapter.) Different icons appear in this area on each Home screen.

 Treat the iPhone screen carefully. It's made of glass and will smudge when you touch it (and will break if you throw it at the wall).

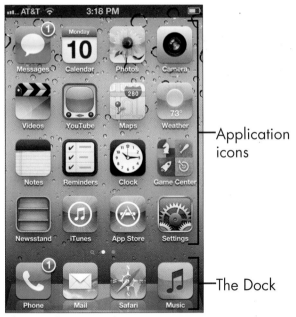

Application icons

The Dock

Figure 2-1

The iPhone uses *touchscreen technology:* When you swipe your finger across the screen or tap it, you're providing input to the device. You hear more about the touchscreen in the next task, but for now, go ahead and play with it for a few minutes — really, you can't hurt anything. Use the pads of your fingertips (not your fingernails) and follow these steps:

1. Tap the Settings icon. The various settings (which you read more about throughout this book) appear, as shown in **Figure** 2-2.

2. To return to the Home screen, press the Home button.

3. Swipe a finger or two from right to left on the Home screen. If downloaded apps fill additional Home screens, then this action moves you to the next Home screen. Note that the little dots at the bottom of the screen, above the Dock icons, indicate which Home screen is displayed. The magnifying glass on the far left represents the Spotlight Search screen.

Figure 2-2

4. To experience the screen rotation feature, hold the iPhone firmly while turning it sideways. The screen flips to the horizontal orientation. To flip the screen back, just turn the device so it's oriented like a pad of paper again. (Note that some apps force iPhone to stay in one orientation or the other.)

 You can customize the Home screen by changing its *wallpaper* (background picture) and brightness. You can read about making these changes in Chapter 7.

Say Goodbye to Click and Drag, Hello to Tap and Swipe

You can use several methods for getting around and getting things done in iPhone using its Multi-Touch screen, including

➠ **Tap once.** To open an application on the Home screen, choose a field such as a search box, select an item in a list, select an arrow to move back or forward one screen, or follow an online link, tap the item once with your finger.

➡ **Tap twice.** Use this method to enlarge or reduce the display of a web page (see Chapter 10 for more about using the *Safari* web browser) or to zoom in or out in the Maps app.

➡ **Pinch.** As an alternative to the tap-twice method, you can pinch your fingers together or move them apart on the screen (see **Figure 2-3**) when you're looking at photos, maps, web pages, or e-mail messages to quickly reduce or enlarge them, respectively.

 You can use the three-finger tap to zoom your screen to be even larger or use multitasking gestures to swipe with 4 or 5 fingers (see the "Explore Multitasking Gestures" task later in this chapter). This method is handy if you have vision challenges. Go to Chapter 7 to discover how to turn on this feature using Accessibility settings.

➡ **Drag to scroll (known as *swiping*).** When you press your finger to the screen and drag to the right or left or drag up or down, the screen moves (see **Figure 2-4**). Swiping to the right on the Home screen, for example, moves you to the *Spotlight* screen (the iPhone search screen). Swiping down while reading an online newspaper moves you down the page; swiping up moves you back up the page.

➡ **Flick.** To scroll more quickly on a page, quickly flick your finger on the screen in the direction you want to move.

➡ **Tap the Status bar.** To move quickly to the top of a list, web page, or e-mail message, tap the Status bar at the top of the iPhone screen.

Figure 2-3

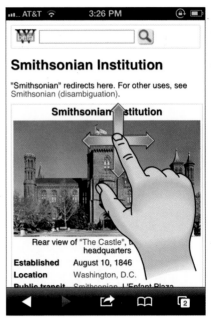

Figure 2-4

→ **Press and hold.** If you're using Notes or Mail or any other application that lets you select text, or if you're on a web page, pressing and holding text selects a word and displays editing tools you can use to select, cut, or copy the text.

Try these methods now by following these steps:

1. Tap the Safari button to display the web browser. (You may be asked to enter your network password to access the network.)

2. Tap a link to move to another page.

3. Double-tap the page to enlarge it; then pinch your fingers together on the screen to reduce its size.

4. Drag one finger around the page to scroll.

5. Flick your finger quickly on the page to scroll more quickly.

6. Press and hold your finger on a word that isn't a link. The word is selected, and the Copy/Define tool is displayed, as shown in **Figure 2-5**. (This step is tricky, so if you don't get it right the first time, don't worry — I cover text editing in more detail in Chapter 5.)

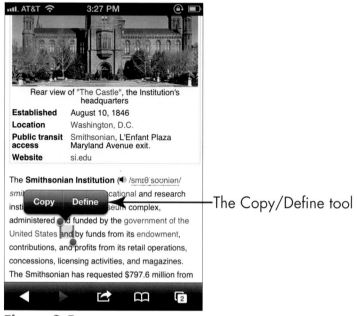

The Copy/Define tool

Figure 2-5

7. Press and hold your finger on a link or an image. A menu appears with commands you select to open the link or picture, open it in a new page, add it to your Reading List (see Chapter 10), or copy it. The image menu also offers the Save Image command. Tap Cancel to close the menu without making a selection.

8. Position your fingers slightly apart on the screen, and then pinch your fingers together to reduce the page; with your fingers already pinched together, place them on the screen, and then move them apart to enlarge the page.

9. Press the Home button to go back to the Home screen.

Display and Use the Onscreen Keyboard

1. The built-in iPhone keyboard appears whenever you're in a text-entry location, such as a search field or an e-mail message. Tap the Notes icon on the Home screen to open this easy-to-use notepad.

2. Tap the note; the onscreen keyboard appears.

3. Type a few words using the keyboard, as shown in **Figure 2-6**. To make the keyboard display as wide as possible, rotate your iPhone to landscape (horizontal) orientation.

4. If you make a mistake while using the keyboard — and you will, when you first use it — press the Delete key (it's in the bottom corner, with the little *x* on it) to delete text to the left of the insertion point.

5. To create a new paragraph, press the Return button, just as you would do on a regular computer keyboard.

6. To type numbers and symbols, press the number key (labeled .?123) on the left side of the spacebar (refer to **Figure 2-6**). The characters on the keyboard change. If you type a number and then tap the spacebar, the keyboard returns to the letter keyboard automatically. To return to the letter keyboard at any time, simply tap one of the letter keys (labeled ABC) on the left side of the spacebar.

7. Use the Shift buttons just as you would on a regular keyboard to type uppercase letters or alternate characters.

8. Double-tap the Shift key to turn on the Caps Lock feature; tap the Shift key once to turn off Caps Lock. (You can control whether this feature is available in iPhone General Settings under Keyboard.)

9. To type a variation on a symbol (for example, to see alternate currency symbols when you press the dollar sign on the number keyboard), hold down the key; a set of alternate symbols appears (see **Figure** 2-7). Note that this trick works with only certain symbols.

10. Tap the Home button to return to the Home screen.

To type a period and space, just double-tap the spacebar.

Note that the small globe symbol you see on the keyboard will only appear if you've enabled multi-language functionality in iPhone settings.

Figure 2-6

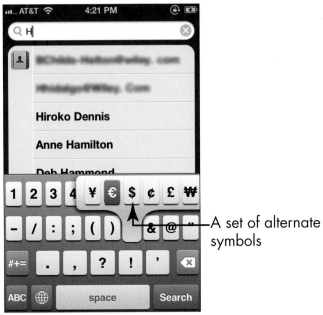

A set of alternate symbols

Figure 2-7

Flick to Search

1. The Spotlight Search feature in iPhone helps you find photos, music, e-mails, contacts, movies, and more. Press and drag from left to right on the Home screen or tap the small magnifying glass symbol farthest to the left at the bottom of the Home screen (but above the Dock) to display the Spotlight screen. (You can also, from the primary Home screen, press the left side of the Home button to move one screen to the left.)

2. Tap in the Search iPhone field (see **Figure 2-8**); the keyboard appears.

3. Begin entering a search term. In the example in **Figure 2-9**, after I typed the letters *No*, the search results displayed some contacts, the Notes app, and a TV show and movie I had downloaded. As you continue to type a search term, the results narrow to match it.

4. Tap an item in the search results to open it.

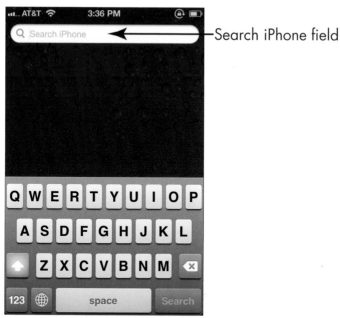

Search iPhone field

Figure 2-8

Figure 2-9

Update the Operating System to iOS 5

1. This book is based on the latest version of the iPhone operating system at the time: iOS 5. To make sure you have the latest and greatest features in iPhone, update to the latest iOS now (and periodically). If you have set up an iCloud account on your iPhone, updates can be set up to happen automatically, or you can use a physical connection to a computer to update the iOS. To use the latter method, plug the 30-pin end of the Dock Connector to USB Cable into your iPhone and plug the USB end into your computer; or if you've set up wireless syncing be within range of a Wi-Fi network.

2. When iTunes opens, click your iPhone (in the Devices section of the Source List on the left side of the computer screen) and then click the Summary tab if it isn't already displayed (see **Figure 2-10**).

3. Read the note next to the Check for Update button to see whether your iOS is up to date. If it isn't, click the Check for Update button. iTunes checks to find the latest iOS version and walks you through the updating procedure.

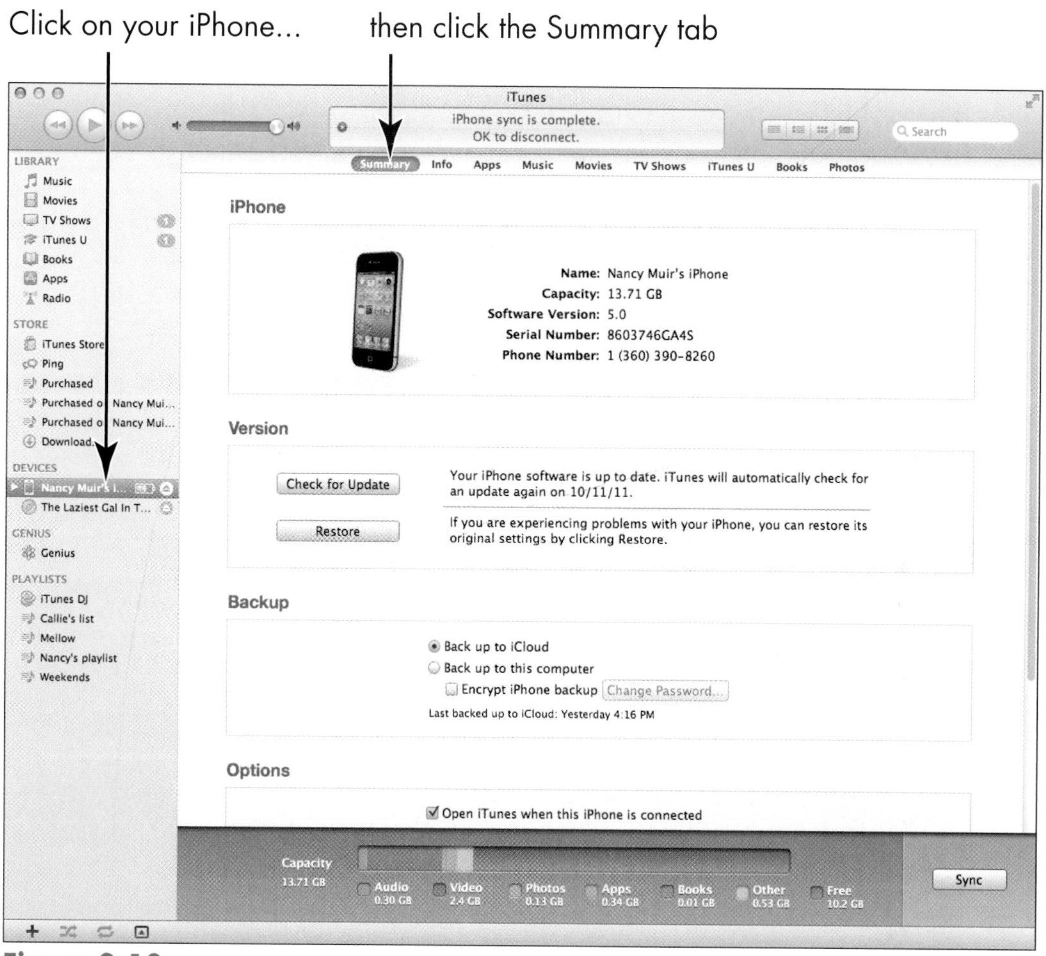

Click on your iPhone...   then click the Summary tab

Figure 2-10

Follow Multitasking Basics

1. *Multitasking* lets you easily switch from one app to another without closing the first one and returning to the Home screen. First, open an app.

2. Double-tap the Home button.

3. On the horizontal bar that appears beneath the Dock at the bottom of the screen (see **Figure 2-11**), flick to scroll to the left or right to locate another app you want to display.

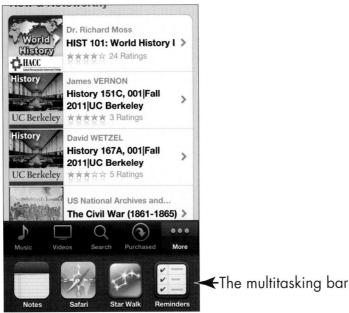

←The multitasking bar

Figure 2-11

4. Tap the app to open it.

 Swiping left to right on the multitasking bar displays the controls for volume and playback, as well as a button that locks and unlocks screen rotation.

Explore Multitasking Gestures

 Multitasking involves jumping from one app to another. There are some gestures you can use with four or five fingers to multitask. Here are the three gestures you can make:

⟹ Swipe up with four or five fingers on any Home screen to reveal the multitasking bar.

⟹ Swipe down with four or five fingers to remove the multitasking bar from the Home screen.

⟹ With an app open, swipe left or right using four or five fingers and you move to another app.

Examine the iPhone Cameras

iPhone 4S has front- and back-facing cameras. You can use the cameras to take still photos (covered in more detail in Chapter 16) or shoot videos (covered in Chapter 17).

For now, take a quick look at your camera by tapping the Camera app icon on the Home screen. The app opens, as shown in **Figure 2-12.**

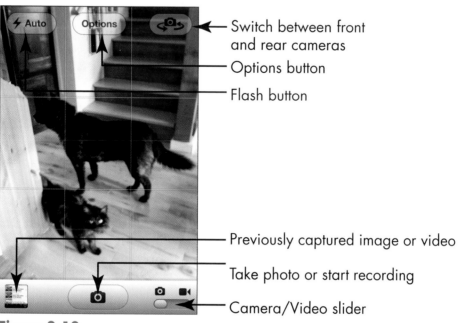

Switch between front and rear cameras

Options button

Flash button

Previously captured image or video

Take photo or start recording

Camera/Video slider

Figure 2-12

You can use the controls on the screen to

➠ Switch between the front and rear cameras.

➠ Change from still-camera to video-camera operation by using the Camera/Video slider.

➠ Take a picture or start recording a video.

➠ Turn on a grid to help you autofocus on still photo subjects.

➠ Turn HDR (high dynamic range for better contrast) on or off.

➠ Tap the Flash button to set flash to On, Off, or Auto.

➠ Open previously captured images or videos.

When you view a photo or video, you can use an iPhone feature to send the image via a tweet or e-mail, print images, use a still photo as wallpaper or assign it to represent a contact, or run a slideshow or edit a video. See Chapters 16 and 17 for more detail about using the iPhone cameras.

Lock Screen Rotation

Sometimes you don't want your screen orientation to flip around when you move your phone around. Use these steps to lock and unlock screen rotation:

1. Double-tap the Home button.

2. Scroll to the left on the multitasking bar that appears along the bottom of the screen.

3. Tap the Orientation button to the far left to toggle between locking and unlocking the orientation in Portrait mode.

Explore the Status Bar

Across the top of the iPhone screen is the *Status bar* (see **Figure 2-13**). Tiny icons in this area can provide useful information, such as the time, battery level, and wireless-connection status. **Table 2-1** lists some of the most common items you find on the Status bar:

Figure 2-13

Table 2-1		Common Status Bar Icons
Icon	**Name**	**What It Indicates**
	Wi-Fi	You're connected to a Wi-Fi network.
	Activity	A task is in progress — a web page is loading, for example.
3:30 PM	Time	You guessed it: You see the time.
	Screen Rotation Lock	The screen is locked and doesn't rotate when you turn the iPhone.
	Play	A media element (such as a song or video) is playing.
79%	Battery Life	The charge percentage remaining in the battery. The indicator changes to a lightning bolt when the battery is charging.

 If you have GPS, 3G, cellular, or Bluetooth service or a connection to a virtual private network (VPN), a corresponding symbol appears on the Status bar whenever one of these features is active. (If you can't even conceive of what a virtual private network is, my advice is not to worry about it.)

Take Inventory of Preinstalled Applications

The iPhone comes with certain functionality and applications — or *apps*, for short — built in. When you look at the Home screen, you see icons for each app. This task gives you an overview of what each app does. (You can find out more about every one of them as you read different chapters in this book.) The icons in the Dock (see the "Meet the Multi-Touch Screen" task, earlier in this chapter) are, from left to right:

⟹ **Phone:** Use this app to make and receive phone calls, view recent calls, create a list of favorite contacts, access your voice mail, and view Contacts.

⟹ **Mail:** You use this application to access e-mail accounts that you have set up in iPhone. Your e-mail is then displayed without your having to browse to the site or sign in. Then you can use tools to move among a few preset mail folders, read and reply to e-mail, and download attached photos to your iPhone. Read more about e-mail accounts in Chapter 11.

⟹ **Safari:** You use the Safari web browser (see **Figure 2-14**) to navigate on the Internet, create and save bookmarks of favorite sites, and add web clips to your Home screen so that you can quickly visit favorite sites from there. You may have used this web browser (or another, such as Internet Explorer) on your desktop computer.

⟹ **Music:** *Music* is the name of your media player. Though its main function is to play music, you can use it to play audio podcasts or audiobooks as well.

Figure 2-14

Apps with icons above the Dock and closer to the top of the Home screen include:

⟶ **Photos:** The photo application in iPhone (see **Figure 2-15**) helps you organize pictures in folders, e-mail photos to others, use a photo as your iPhone wallpaper, and assign pictures to contact records. You can also run slideshows of your photos, open albums, pinch or unpinch to shrink or expand photos, and scroll photos with a simple swipe.

⟶ **Calendar:** Use this handy onscreen daybook to set up appointments and send alerts to remind you about them.

⟶ **Contacts:** Tucked into the Utilities folder, in the address book feature (see **Figure 2-16**), you can enter contact information (including photos, if you like, from your Photos or Cameras app) and share contact information by e-mail.

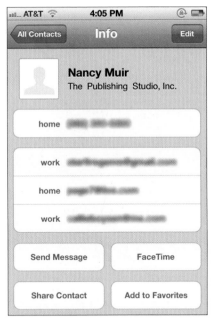

Figure 2-15   Figure 2-16

 Notes: Enter text or cut and paste text from a website into this simple notepad app. You can't do much except save notes or e-mail them — the app has no features for formatting text or inserting objects. You'll find Notes handy, though, for simple notes on the fly.

 Maps: In this cool iPhone version of Google Earth, you can view classic maps or aerial views of addresses, find directions from one place to another by car, foot, or public transportation, and even view some addresses as though you're standing in front of the building at street level (though not every street image is totally up to date).

 Videos: This media player is similar to Music but specializes in playing videos and offers a few features specific to this type of media, such as chapter breakdowns and information about a movie's plot and cast.

➠ **YouTube:** Tap this icon to go to the popular online video-sharing site, where you can watch videos that people have posted and then comment on the videos or share them with others.

➠ **iTunes:** Tapping this icon takes you to the iTunes store, where you can shop 'til you drop (or until your iPhone battery runs out of juice) for music, movies, TV shows, audiobooks, and podcasts and then download them directly to your iPhone. (See Chapter 12 for more about how iTunes works.)

➠ **App Store:** Here, you can buy and download applications that do everything from enabling you to play games to building business presentations. You can also subscribe to periodicals for use with the Newsstand app, described later. Some of these are even free!

➠ **Settings:** This isn't exactly an app, but it's an icon you should know about, anyway: It's the central location on the iPhone where you can specify settings for various functions and do administrative tasks such as set up e-mail accounts or create a password.

➠ **Game Center:** This app helps you browse games in the App Store and play them with other people online. You can add friends and track your scores. See Chapter 18 for more about Game Center.

➠ **Camera:** As you may have read earlier in this chapter, the Camera app is control central for the still and video cameras built into the iPhone. Camera is also the app you use to make and receive FaceTime (video) calls.

Newsstand: Similar to an e-reader for books, Newsstand is a handy reader app for subscribing to and reading magazines, newspapers, and other periodicals.

 Reminders: This is a useful app that centralizes all your calendar entries and alerts to keep you on schedule, as well as allowing you to create to-do lists.

➡ **Messages:** For those who love to instant message, the Messages app comes to the rescue. iMessage brings more features to the Messages app that has been in iPhone for quite some time. Now you can engage in live text- and image-based conversations with others via their phones or other devices that use e-mail.

 Tucked in the Utilities folder are some handy tools: Calculator, Compass, and Voice Memo apps. The Utilities folder is on the second home screen by default.

 The iBooks application isn't bundled with the iPhone out of the box. Though iBooks is free, you have to download it from the App Store. Because the iPhone has been touted as an outstanding *e-reader* — a device that enables you to read books on an electronic device, similar to the Amazon Kindle — you should definitely consider downloading the app as soon as possible. (For more about downloading applications for your iPhone, see Chapter 13. To work with the iBooks e-reader application itself, go to Chapter 14.)

Lock iPhone, Turn It Off, or Unlock It

Earlier in this chapter, I mention how simple it is to turn on the power to your iPhone. Now it's time to put it to *sleep* (a state in which the screen goes black, though you can quickly wake up the iPhone) or turn off the power to give your new toy a rest. Here are the procedures you can use:

➠ **Press the On/Off button.** The iPhone goes to sleep: The screen goes black and is locked.

➠ **Press and hold the On/Off button until the Slide to Power Off bar appears at the top of the screen, and then swipe the bar.** You've just turned off your iPhone.

➠ **Press the Home button and swipe the onscreen arrow on the Slide to Unlock bar (see Figure 2-17).** The iPhone unlocks.

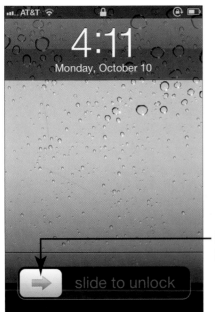

Slide this button
to unlock the phone

Figure 2-17

 The iPhone automatically enters sleep mode after a few minutes of inactivity. You can change the time interval at which it sleeps by adjusting the Auto-Lock feature in Settings. See this book's companion Cheat Sheet at www.dummies.com/cheatsheet/ iphoneforseniors to review tables of various settings.

Getting Going

*Y*our first step in getting to work with the iPhone is to make sure that its battery is charged. Next, if you want to find free or paid content for your iPhone from Apple, from movies to music to e-books to audiobooks, look into opening an iTunes account.

After that, connect your iPhone to your computer and sync them so that you can exchange content between them (for example, to transfer your saved photos or music to the iPhone).

If you prefer, you can take advantage of the new iCloud service from Apple to store and push all kinds of content and data to all your Apple devices — wirelessly.

This chapter also introduces you to the *iPhone User Guide*, which you access using the Safari browser on your iPhone. The guide essentially serves as your iPhone Help system, to provide advice and information about your magical new device.

Charge the Battery

1. My iPhone showed up in the box fully charged, and let's hope yours did, too. Because all batteries run down eventually, one of your first priorities is to know how to recharge your iPhone battery. Gather your iPhone and its connector cord and power adapter.

Chapter 3

Get ready to . . .

2. Gently plug the USB end (the smaller of the two connectors) of the Dock Connector to USB Cable into the USB Power Adapter.

Note that if you have a hard case for your iPhone remove the phone from it while charging as these cases retain heat which is bad for the phone and case.

3. Plug the other end of the cord into the cord connector slot on the iPhone (see **Figure 3-1**).

Attach the USB connector...   to the power adapter.

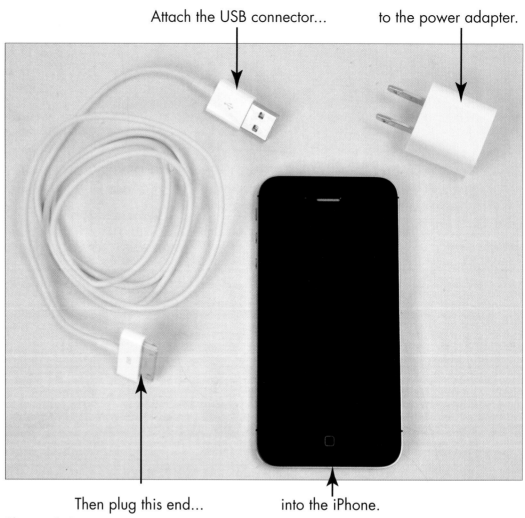

Then plug this end...   into the iPhone.

Figure 3-1

4. Unfold the two metal prongs on the power adapter (refer to **Figure 3-1**) so they extend from it at a 90-degree angle, and then plug the adapter into an electric outlet.

 If you buy the Apple iPhone Dock accessory, you can charge your iPhone while it's resting in the Dock. Just plug the larger connector into the back of the Dock instead of at the bottom of the iPhone.

Download iTunes to Your Computer

1. If you're using a Mac, you already have iTunes installed but to be sure you have the latest version click iTunes, and choose Check for Updates. For Windows users, you should download the iTunes application to your computer so that you have the option of using it to *sync* (transfer) downloaded content to your iPhone. Go to www.apple.com/itunes using your computer's browser.

2. Click the iTunes Free Download link (see **Figure 3-2**). On the screen that opens, click the Download Now button.

3. In the dialog that appears (see **Figure 3-3**), or using the Internet Explorer 9 download bar if you have that version, click Run. The iTunes application downloads.

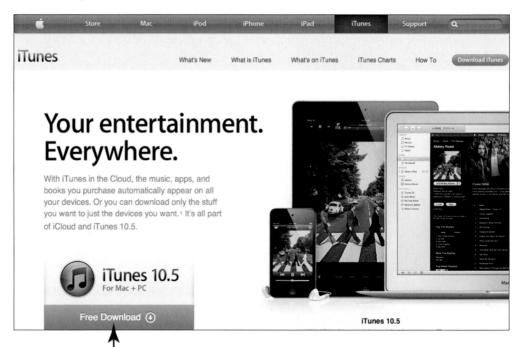

Click this button

Figure 3-2

Click Run

Figure 3-3

4. When the download is complete, another dialog appears, asking whether you want to run the software. Click Run, and the iTunes Installer appears (see **Figure** 3-4).

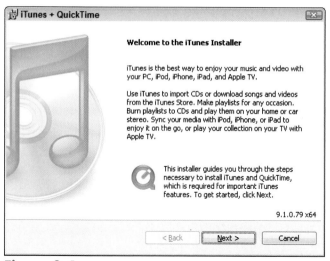

Figure 3-4

5. Click Next.

6. Click the I Accept the Terms of the License Agreement check box in the following dialog and click Next.

7. Review the installation options, click to deselect the ones you don't want to use, and then click the Install button, as shown in **Figure 3-5**. A dialog appears, showing the installation progress.

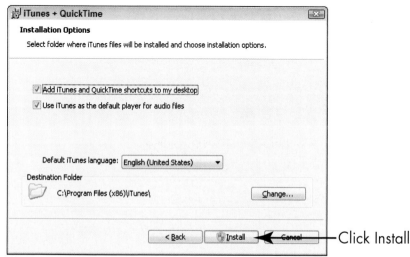

Click Install

Figure 3-5

8. When a dialog appears and tells you that the installation is complete, click Finish. You have to restart your computer for the configuration settings that were made during the installation to take effect.

 As of this writing, iTunes 10.5 is the latest version of this software. It includes a paid feature called iTunes Match that allows you to store all your music, even music that you've copied from CDs, in the cloud and push it to all your Apple devices using iCloud.

Open an iTunes Account for Music, Movies, and More

1. To be able to buy or download free items from iTunes or the App Store on your computer or iPhone, you must open an iTunes account. First, open the iTunes app (you download it to your computer in the preceding task). You can open iTunes by clicking its option on your computer's Start menu in Windows or by clicking the iTunes item in the Mac Dock or Launchpad.

2. Click the Store tab to open the Store menu and choose Create Account from the menu that appears (see **Figure 3-6**).

Figure 3-6

3. On the Welcome to the iTunes Store screen that appears, click Continue.

4. On the following screen (see **Figure 3-7**), click to select the I Have Read and Agree to the iTunes Terms and Conditions check box, and then click the Continue button.

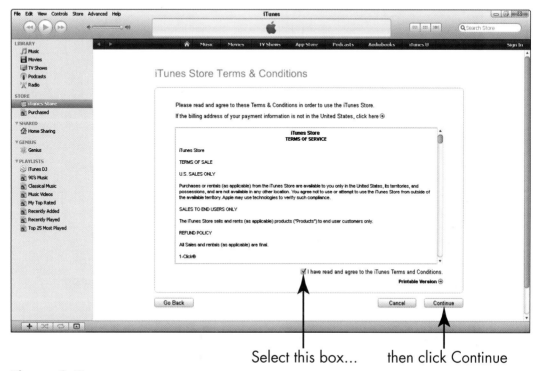

Select this box...   then click Continue

Figure 3-7

5. In the Create iTunes Store Account (Apple ID) screen that follows (see **Figure 3-8**), fill in the information fields, click the last two check boxes to deselect them if you don't want to receive e-mail from Apple, and then click the Continue button.

6. On the Provide a Payment Method screen that appears (see **Figure 3-9**), enter your payment information and click the Continue button.

Figure 3-8

Figure 3-9

7. A screen appears, confirming that your account has been opened. Click the Done button to return to the iTunes Store.

 If you prefer not to leave your credit card info with Apple, buy an iTunes gift card and provide that as your payment information. You can replenish the card perioidically through the Apple Store.

Make iPhone Settings Using iTunes

1. Open your iTunes software. (On a Windows computer, choose Start⇨All Programs⇨iTunes; on a Mac, click the iTunes icon in the Dock or Launchpad.)

2. iTunes opens and, if you've connected it to your computer using the Dock Connector to USB Cable, your iPhone is listed in the Devices section of the Source List on the left, as shown in **Figure 3-10.** Click on your iPhone, and a series of tabs displays. The tabs offer information about your iPhone and settings to determine how to download music, movies, or podcasts, for example. (You can see the simple choices on the Music tab in **Figure 3-11.**) The settings relate to the kind of content you want to download and whether you want to download it automatically (when you sync) or manually. See **Table 3-1** for an overview of the settings that are available on each tab.

Click on your iPhone...   to display this series of tabs

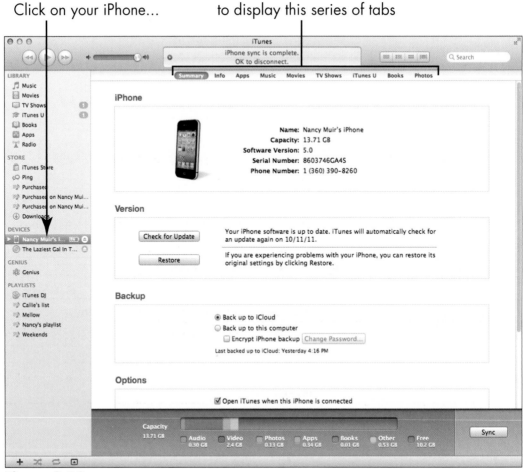

Figure 3-10

Figure 3-11

3. Make all settings for the types of content you plan to
obtain on your computer. **Table 3-1** provides information about settings on the different tabs.

Table 3-1	iPhone Settings in iTunes
Tab Name	**What You Can Do with the Settings on the Tab**
Summary	Perform updates to the iPhone software and set general syncing options.
Info	Specify which information to sync: Contacts, Calendars, E-mail accounts, Bookmarks, or Notes.
Apps	Sync with iPhone the apps you've downloaded to your computer and manage the location of those apps and folders, as well as associating data files with associated apps.
Music	Choose which music to download to your iPhone when you sync.
Movies	Specify whether to automatically download movies and which movies to sync.
TV Shows	Choose shows and episodes to sync automatically.
iTunes U	Choose free online courses to download.
Podcasts	Choose podcasts and episodes to sync automatically.
Books	Choose to sync all, or only selected, books and audiobooks to your iPhone.
Photos	Choose the applications or folders from which you want to download photos or albums.

Sync the iPhone to Your Computer Using iTunes

1. After you specify which content to download in iTunes (see the preceding task), you can use the Dock Connector to USB Cable to connect your iPhone and computer at any time and sync files, contacts, calendar settings, and more. You can also use the iTunes Wi-Fi Sync setting to allow cordless syncing (Tap Settings, General, iTunes Wi-Fi Sync, and then tap Sync Now to sync with a computer connected to the same Wi-Fi network). After iTunes is downloaded to your computer and your iTunes account is set up, plug the data connection cord into your iPhone (using the wider connector).

2. Plug the other end of the cord into a USB port on your computer.

3. iTunes opens and shows an item for your iPhone in the Source List on the left (refer to **Figure 3-10**). Your iPhone screen shows the phrase *Sync in Progress.*

4. When the syncing is complete, the Lock screen returns on the iPhone; iTunes shows a message above the series of tabs (refer to **Figure 3-10**), indicating that the iPhone sync is complete and that you can disconnect the cable. Any media you chose to transfer in your iTunes settings, and any new photos in the photos folder on your computer, have been transferred to your iPhone.

Understand iCloud

There's an option to syncing content by using iTunes. Concurrent with the launch of iOS 5, iCloud is a service offered by Apple that allows you to back up all your content to online storage. That content is then pushed automatically to all your Apple devices through a wireless connection. All you need to do is get an iCloud account, which is free, and make settings on each device and in iTunes for which types of content you want pushed to each device. After you've done that, any content you create or purchase on one device — such as music, apps, and TV shows, as well as documents created in Apple's iWork apps, photos, and so on — is synced among your devices automatically.

When you get an iCloud account, you get 5GB of free storage; content you purchase (such as apps, books, music, iTunes Match content, Photo Stream contents, and TV shows) won't be counted against your storage. If you want additional storage, you can buy an upgrade from one of your devices. 10GB costs $20 per year; 20GB is $40; and 50GB is $100 per year. Most people will do just fine with the free 5GB of storage.

To upgrade your storage, go to iCloud in Settings, tap Storage & Backup, and then tap Buy More Storage. Tap the amount you need and then tap Buy.

You can make settings for backing up your content to iCloud in the iCloud section of General Settings. You can have content backed up automatically, or do it manually. See Chapter 23 for more about this topic.

Get an iCloud Account

Before you can use iCloud, you need an iCloud account, which is tied into the Apple ID you probably already have. You can turn on iCloud when first setting up your iPhone or use Settings to sign up using your Apple ID.

1. When first setting up your phone after upgrading to iOS 5, in the sequence of screens that appear you'll see the one in **Figure 3-12**. Tap Use iCloud.

— Tap this option

Figure 3-12

2. In the next dialog, tap Backup to iCloud. Your account is now set up based on the Apple ID you entered earlier in the setup sequence.

Here are the steps to set up iCloud on your iPhone if you didn't do so when first setting up iPhone:

1. Tap Settings and then tap iCloud.

2. Tap the On/Off button to turn on iCloud.

3. Enter your Apple ID and password and tap the Sign In button. (See **Figure 3-13**.) (If you don't have an Apple ID, tap the Get a Free Apple ID button and follow the instructions to get your ID.) A dialog appears, asking whether you'd like to merge your iPhone calendars, reminders, and bookmarks with iCloud.

Figure 3-13

4. A dialog may appear asking if you want to allow iCloud to use the location of your iPhone. Tap OK. Your account is now set up.

Make iCloud Sync Settings

1. When you have an iCloud account up and running (see the previous task), you have to specify which type of content should be synced with your iPhone via iCloud. To do so, tap Settings and then tap iCloud.

2. In the iCloud settings shown in **Figure** 3-14, tap the On/Off button for any item that's turned off that you want to turn on (or vice versa). You can sync Mail, Contacts, Calendars, Reminders, Bookmarks, and Notes.

Select the content to sync via iCloud

Figure 3-14

3. To turn Photo Stream, Documents & Data, or Storage & Backup on or off (so you can sync photos, documents created in iWork, or settings data, respectively), tap those options on the list (refer to **Figure** 3-14) and then tap the On/Off button for each particular setting in the subsequent screen.

If you want to allow iCloud to provide a service for locating a lost or stolen iPhone, tap the On/Off button in the Find My iPhone field to activate it. This service helps you locate, send a message to, or delete content from your iPhone if it falls into other hands.

View the iPhone User Guide Online

1. The *iPhone User Guide* is equivalent to the Help system you may have used on a Windows or Mac computer. You access the guide online as a bookmarked site in the Safari browser. From the iPhone Home screen, tap the Safari icon.

 2. Tap the Bookmark icon. On the Bookmarks menu that appears (see **Figure** 3-15), tap iPhone User Guide.

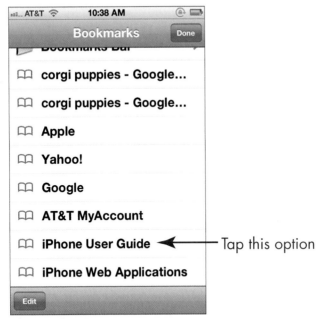

Figure 3-15

3. Tap a topic (see the image on the left in **Figure** 3-16) to display subtopics (see the image on the right in **Figure** 3-16).

Tap a topic...   to display its subtopics

Figure 3-16

4. Tap a subtopic to display information about it, as shown in **Figure** 3-17.

5. Tap any link in the subtopic information to access additional topics.

Figure 3-17

6. Tap the Home screen button to close the browser.

 To find the PDF version of the *iPhone User Guide* on your browser, see `http://support.apple.com/ manuals/#iphone` and select the User Guide appropriate for your iPhone model and iOS.

Part II
Start Using Your iPhone

The 5th Wave — By Rich Tennant

"Hold on Barbara. I'm pretty sure there's an app for this."

Making and Receiving Calls

*I*f you're the type who wants a cellphone only to make and receive calls, you probably didn't buy an iPhone. Still, making and receiving calls is one of the main functions of any phone, smart or not.

In this chapter, you discover all the basics of placing calls, receiving calls, and using available tools during a call to mute the call, turn on speakerphone, and more.

Use the Keypad to Place a Call

1. Dialing a call with a keypad is an obvious first skill for you to acquire, and it's dead simple. On any Home screen, tap Phone in the Dock, and the keypad appears (see **Figure 4-1**). (Note that if anything other than the keypad appears, you can just tap the Keypad button at the bottom of the screen to display the keypad.)

2. Enter the number you want to call by tapping the number buttons; as you do, the number appears at the top of the keypad.

3. If you enter a number incorrectly, use the Delete button on the keypad (a backward-pointing arrow with an X in it) to clear numbers one at a time.

4. Tap Call. The call is placed, and tools appear as shown in
Figure 4-2.

Add Contact button

Delete button

Keypad button

Figure 4-1

Figure 4-2

When you enter a phone number, before you place the call you can tap the Add Contact button to the left of the Call button to add the person to your Contacts app. You can create a new contact or add the phone number to an existing contact using this feature.

 If you're on a call that requires that you punch in numbers or symbols such as a pound sign, tap the Keypad button on the tools that appear during a call to display the keypad. See more about using calling tools in the last task of this chapter.

End a Call

In the following tasks, I tell you several other ways to place calls; however, I don't want to leave you on your first call without a way out. When you're on a phone call, the Call button changes to a red End button (see **Figure** 4-3). Tap End, and the call is disconnected.

Figure 4-3

Place a Call Using Contacts

1. If you've created a contact and included a phone number in that contact record, you can use the Contacts app to place a call. Tap the Phone icon in any Home screen dock.

2. Tap the Contacts button at the bottom of the screen.

3. In the Contacts list that appears (see **Figure** 4-4), scroll up or down to locate the contact you need or tap a letter along the right side to jump to that section of the list.

4. Tap the contact to display his or her record. In the record that appears (see **Figure** 4-5), tap the phone number field. The call is placed.

Contacts button

Figure 4-4

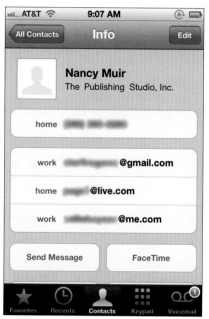

Figure 4-5

 If you locate a contact and the record doesn't include a phone number, you can add it at this point by tapping the Edit button, entering the number, and tapping Done. Then place your call following Step 4 in the preceding steps list.

Return a Recent Call

1. If you want to dial a number from a call you've recently made or received, you can use Recents. Tap Phone in the Dock on any Home screen.

2. Tap the Recents button at the bottom of the screen. A list of recent calls that you've both made and received appears (see **Figure** 4-6).

3. If you want to view only the calls you've missed, tap the Missed tab at the top of the screen.

4. Tap the arrow to the right of any item to view information about calls to or from this person (see **Figure** 4-7).

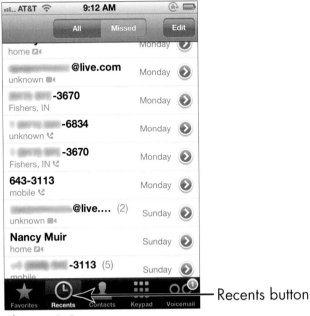

——— Recents button

Figure 4-6

Figure 4-7

5. Tap the Recents button to return to the Recents list, and
then tap any call to place a call to that number.

 To delete calls from your Recents list, with it displayed tap Edit. The list appears with circles containing minus symbols to the left of each item. Tap one
of these circles and then tap the Delete button that
appears. Tap Done to return to the Recents list.

Use Favorites

1. You can save contacts to Favorites in the Phone app so
you can quickly make calls to your A-list folks or businesses. Tap Phone on any Home screen.

2. Tap the Favorites button at the bottom of the screen.

3. In the Favorites screen that displays (see **Figure** 4-8), tap the Add button.

4. Your Contacts list appears. Locate a contact you want to make a Favorite, and tap it. In the contact record, tap the phone number field. A menu appears (see **Figure** 4-9).

5. Tap Voice Call or FaceTime depending on which type of call you prefer to make to this person most of the time. The Favorites list reappears with your new favorite contact on it.

6. To place a call to a Favorite, display the Phone app, tap Favorites, and then tap a person on the list to place a call.

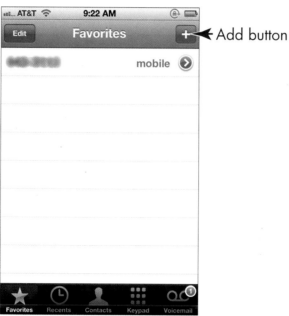

Figure 4-8

Figure 4-9

 If you decide to place a FaceTime call to a Favorite you've created using the Voice Call setting, just tap the arrow to the right of the favorite's listing and tap the FaceTime button in the contact record that

appears. You can also create two contacts for the
same person, one with a cellphone and one with a
land phone, for instance, and place one or both in
Favorites. See Chapter 9 for more about making
FaceTime calls.

Receive a Call

There's one step to receiving a call. When a call comes in to you, the
screen shown in **Figure 4-10** appears. Tap Answer to pick up the call,
or Decline to end the call without picking up and send the caller into
your voice mail. If you prefer to have the call go to your voice mail,
don't press either button.

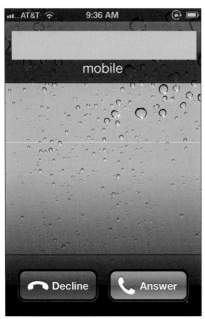

Figure 4-10

Use Tools During a Call

When you're on a call, whether you initiated it or received it, a set of
tools, shown in **Figure 4-11**, is displayed.

Here's what these six buttons allow you to do, starting with the top-left corner:

➡ **Mute:** Silences the phone call so the caller can't hear you. The Mute button turns blue, as shown in **Figure 4-12,** when a call is muted. Tap again to unmute the call.

➡ **Keypad:** Displays the numeric keypad.

➡ **Speaker:** Turns the speakerphone feature on and off.

➡ **Add Call:** Displays Contacts so you can add a caller to a conference call.

➡ **FaceTime:** Begins a video call with somebody who has an iPhone 4 or 4S, iPod touch (4th generation or later), iPad 2, or a Mac.

➡ **Contacts:** Displays a list of contacts.

Figure 4-11

Figure 4-12

Managing Contacts

*C*ontacts is the iPhone equivalent of the dog-eared address book that sits by your phone. The Contacts app is simple to set up and use, and it has some powerful features beyond simply storing names, addresses, and phone numbers.

For example, you can pinpoint a contact's address in iPhone's Maps application. You can use your contacts to address e-mail and Facebook messages and Twitter tweets quickly. If you store a contact record that includes a website, you can use a link in Contacts to view that website instantly. And, of course, you can easily search for a contact by a variety of criteria, including how people are related to you, such as by family ties or mutual friends.

In this chapter, you discover the various features of Contacts, including how to save yourself from having to spend time entering contact information by syncing a contacts list from services such as Google or Yahoo! to your iPhone.

Chapter 5

Add a Contact

1. Scroll to your second Home screen and tap the Utilities icon, and then tap the Contacts app icon to open the application. An alphabetical list of contacts appears, like the one shown in **Figure 5-1**.

2. Tap the Add button, the button with the small plus sign (+) on it. A blank Info page opens (see **Figure 5-2**). Tap in any field, and the onscreen keyboard is displayed.

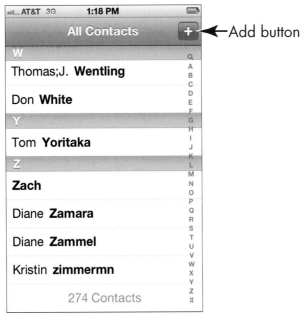

Figure 5-1

3. Enter any contact information you want. (Only one of the First, Last, or Company fields is required.)

Figure 5-2

4. To scroll down the contact page and see more fields, flick up on the page with your finger.

5. If you want to add a mailing or street address, you can tap Add New Address, which opens additional entry fields.

6. To add an information field such as Nickname or Job Title, tap Add Field. In the Add Field dialog that appears (see **Figure 5-3**), choose a field to add. (You may have to flick the page up with your finger to view all the fields.)

7. Tap the Done button when you finish making entries. The new contact appears in your address book. Tap it to see details (see **Figure 5-4**.)

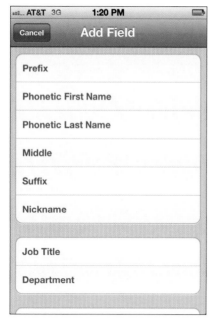

Figure 5-3

Figure 5-4

 If your contact has a name that's difficult for you to pronounce, consider adding the Phonetic First Name or Phonetic Last Name field, or both, to that person's record (refer to Step 6).

Sync Contacts Using iTunes

1. You can use your iTunes account, accessed from your computer, to sync contacts between e-mail accounts such as Yahoo! or Gmail (like the one shown in **Figure 5-5**), for example, and your iPhone Contacts application. This sync works in both directions: Contacts from the iPhone are sent to your e-mail account, and contacts from your e-mail account are sent to the iPhone. First connect your iPhone to your computer using the Dock Connector to USB Cable.

2. In the iTunes application that opens on your computer, click the name of your iPhone (such as Nancy's iPhone), which is now listed in the left navigation area of iTunes.

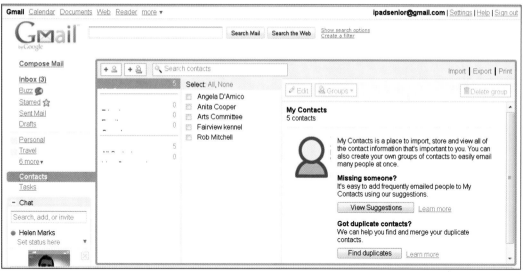

Figure 5-5

3. Click the Info tab, shown in **Figure 5-6**.

4. Click to select the Sync Contacts check box, and then select your e-mail provider.

5. Enter your account information in the dialog that appears.

6. Click the Sync button, and your iPhone screen changes to show that syncing is in progress.

7. When the sync is complete, open Contacts on your iPhone. All contacts have been brought over to it.

8. Unplug the Dock Connector to USB Cable.

 You can also use iCloud to automatically sync contacts among all your Apple devices. See Chapter 3 for more about making iCloud settings to choose whether your contacts are synced through the cloud.

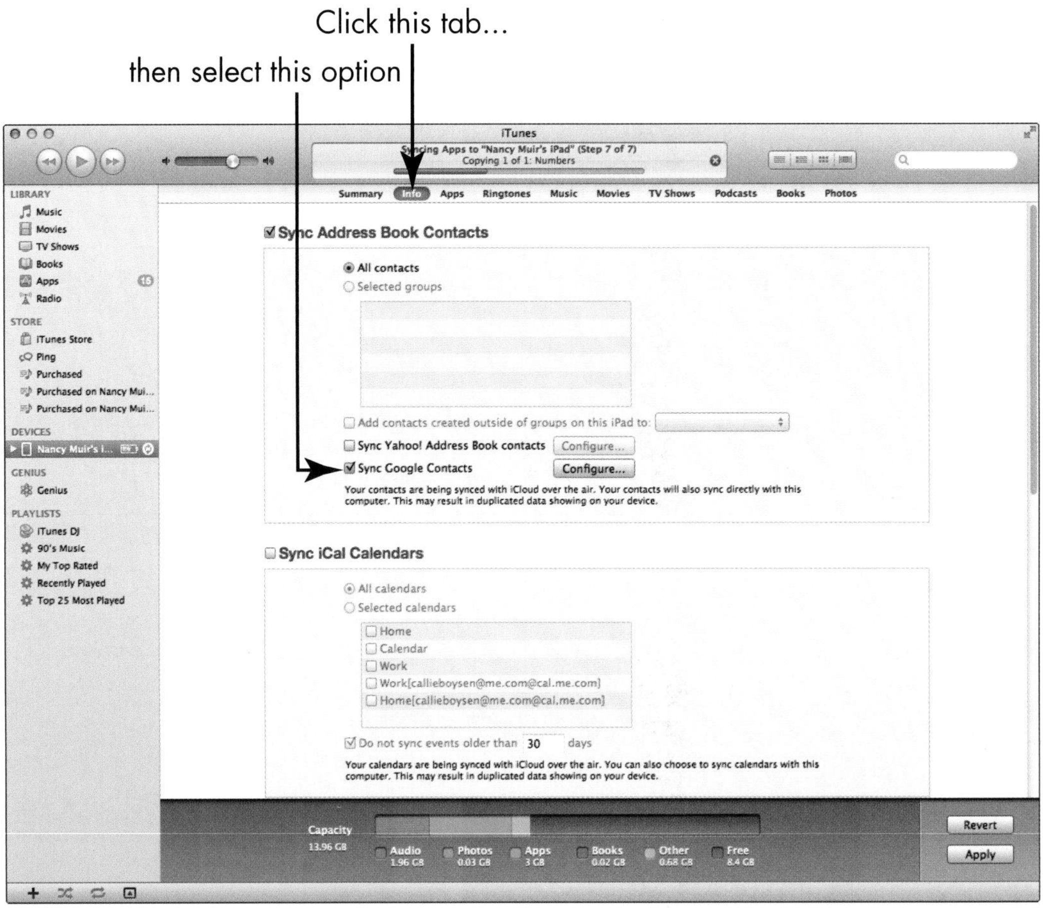

Click this tab...

then select this option

Figure 5-6

> You can also use the iTunes Wi-Fi sync feature in iPhone General Settings to sync with iTunes wirelessly from a computer connected to the same Wi-Fi network.

Assign a Photo to a Contact

1. With Contacts open, tap a contact to whose record you want to add a photo.

2. Tap the Edit button.

3. On the Info page that appears (see **Figure** 5-7), tap Add Photo.

4. In the dialog that appears, tap Choose Photo to choose an existing photo in the Photos app's Camera Roll. You could also choose Take Photo to take that contact's photo on the spot.

5. In the Photos dialog that appears, choose a source for your photo (such as Camera Roll or Photo Stream).

6. In the photo album that appears, tap a photo to select it. The Move and Scale dialog, shown in **Figure** 5-8, appears.

Tap here to add a photo

Figure 5-7   Figure 5-8

7. Tap the Choose button to use the photo for this contact.

8. Tap Done to save changes to the contact. The photo appears on the contact's Info page (see **Figure** 5-9).

Figure 5-9

 While in the Photos dialog in Step 6, you can modify the photo before saving it to the contact information. You can unpinch to expand the photo and move it around the space to focus on a particular section and then tap the Choose button to use the modified version.

Add Twitter or Facebook Information

1. New with iOS 5, you can add Twitter information so you can quickly tweet (send a short message to) others using Twitter. You can also add Facebook information so you can post a message to your contact's Facebook account. With Contacts open, tap a contact.

2. Tap the Edit button.

3. Scroll down and tap Add Field.

4. In the list that appears (see **Figure 5-10**) tap Twitter.

5. A Twitter field opens. Tap in the field and enter the contact's username information for an account.

 If you prefer to add Facebook information instead of Twitter, tap the word *Twitter* on the left side of the Twitter field in Step 5. In the Services dialog that opens, tap Facebook and enter the contact's Facebook information. (You can also choose Flickr, LinkedIn, Myspace, or Add Custom Service in the Services dialog.)

6. Tap Done, and the information is saved. The account is now displayed when you select the contact, and you can send a tweet or Facebook message by simply tapping the username, then tapping the service you want to use to contact them, and then tapping the appropriate command (such as Tweet, as shown in **Figure 5-11**).

Figure 5-10

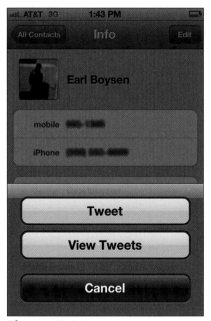

Figure 5-11

Designate Related People

1. You can quickly designate family relations in a contact record if those people are saved to Contacts. One great use for this is using the Siri app to simply say "Call Mom" to call someone who is designated in your contact information as your mother. Tap a contact and then tap Edit.

2. Scroll down the record and tap Add Field.

3. Tap Related People. A new field labeled Mother (see **Figure 5-12**) now appears.

4. Tap the word *mother* and a list of other possible relations appears. Tap one to change the label if you wish.

5. Tap the blue arrow in the field, and your contact list appears. Tap the person's name, and it appears in the field. A new blank field also appears (see **Figure 5-13**).

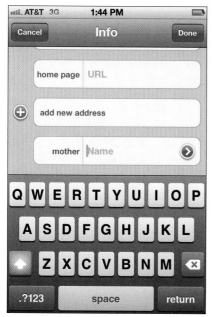

Figure 5-12

Figure 5-13

 After you add relations to a contact record, when you select the person in the Contacts main screen, all the related people for that contact are listed there.

Set Ringtones and Text Tones

1. If you want to hear a unique tone when you receive a phone or FaceTime call from a particular contact, you can set this up in Contacts. For example, if you want to be sure you know instantly if your spouse, sick friend, or boss is calling, set a unique tone for that person. Tap to add a new contact or select a contact in the list of contacts and tap Edit.

2. Tap the Ringtone field, and a list of tones appears (see **Figure 5-14**).

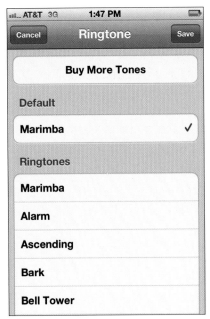

Figure 5-14

3. Tap a tone, and it previews. When you hear one you like, tap Save.

4. Tap Done to save the new tone setting.

 If you set a custom tone for someone that tone will be used when they call or FaceTime you. You can also set a custom text tone to be used when they send you a text message; tap Text Tone instead of Ringtone in Step 2 above and follow the remaining steps.

 If your Apple devices are synced via iCloud, setting a unique ringtone for an iPhone contact will also set it for your iPad. See Chapter 3 for more about iCloud.

Search for a Contact

1. With Contacts open, tap in the Search field at the very top of your contacts list (see **Figure 5-15**). The onscreen keyboard opens.

2. Type the first letter of either the first or last name or company; all matching results appear, as shown in **Figure 5-16**. In the example, pressing *N* displays *American Health Network, Nancy Fletcher, Nolan Lichti, and Nancy Muir* in the results, all of which have *N* as the first letter of the first or last part of the name.

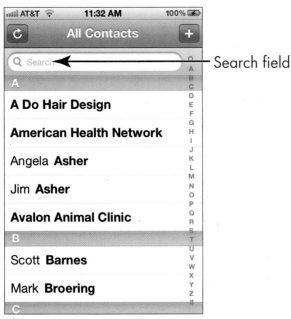

Search field

Figure 5-15

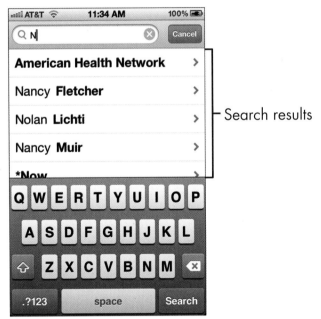

Search results

Figure 5-16

3. Tap a contact in the results to display that person's Info page.

> You can't search by phone number, website, or address in Contacts at the time of this writing, though you can search by these criteria using Spotlight Search. We can only hope that Apple adds this functionality in future versions of the app!

> You can also use the alphabetical listing to locate a contact. Tap and drag to scroll down the list of contacts on the All Contacts page on the left. You can also tap on any tabbed letter along the left side of the page to scroll quickly to the entries starting with that letter.

Go to a Contact's Website

1. If you entered website information in the Home Page field, it automatically becomes a link in the contact's record. Tap the Contacts app icon on the Home screen to open Contacts.

2. Tap a contact's name to display the person's contact information, locate the URL field, and then tap the link (see **Figure 5-17**).

3. The Safari browser opens with the web page displayed (see **Figure 5-18**).

Tap this link

Figure 5-17

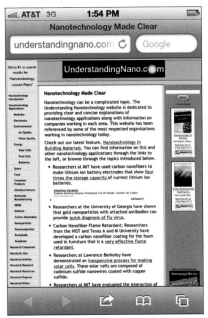

Figure 5-18

You can't go directly back to Contacts after you follow a link to a website. You have to tap the Home button and then tap the Contacts app icon again to re-enter the application or use the multitasking feature by double-tapping the Home button and choosing Contacts from the icons that appear along the bottom of the screen.

Address E-mail Using Contacts

1. If you entered an e-mail address for a contact, the address automatically becomes a link in the contact's record. Tap the Contacts app icon on the Home screen to open Contacts.

2. Tap a contact's name to display the person's contact information, and then tap the e-mail address link labeled Home (see **Figure 5-19**).

3. The New Message dialog appears, as shown in **Figure 5-20**. Initially the title bar of this dialog reads *New*

Message, but as you type a subject, *New Message* changes to the specific title.

Figure 5-19

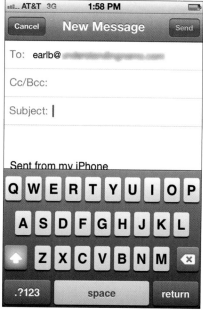

Figure 5-20

4. Use the onscreen keyboard to enter a subject and message.

5. Tap the Send button. The message goes on its way!

Share a Contact

1. After you've entered contact information, you can share it with others via an e-mail message. With Contacts open, tap a contact name to display its information.

2. On the information page, scroll down and tap the Share Contact button. In the dialog that appears, tap Email. A New Message form appears.

 You can also share a contact using a text message. Tap the Share Contact button and then tap Message. The Messages app opens to a New Message screen;

the contact you're sharing already appears in the message's body, so all you have to do is add a recipient in the To field and tap Send.

3. In the New Message form, shown in **Figure 5-21,** use the onscreen keyboard to enter the recipient's e-mail address. Note if the person is saved in Contacts, just type his or her name here.

Figure 5-21

4. Enter information in the Subject field.

5. If you like, enter a message and then tap the Send button. The message goes to your recipient with the contact information attached as a .vcf file. (This *vCard* format is commonly used to transmit contact information.)

When somebody receives a vCard containing contact information, he or she needs only to click the attached file to open it. At this point, depending on the e-mail or contact management program, the recipient can perform various actions to save the content. Other iPhone, iPod touch, or iPhone users can easily import .vcf records into their own Contacts apps.

View a Contact's Location in Maps

1. If you've entered a person's address in Contacts, you have a shortcut for viewing that person's location in the Maps application. Tap the Contacts app icon on the Home screen to open it.

2. Tap the contact you want to view to display his information.

3. Tap the address. Maps opens and displays a map to the address (see **Figure 5-22**).

Figure 5-22

 This task works with more than your friends' addresses. You can save information for your favorite restaurant or movie theater or any other location and use Contacts to jump to the associated website in the Safari browser or to the address in Maps. For more about using Safari, see Chapter 10. For more about the Maps application, see Chapter 19.

Delete a Contact

1. When it's time to remove a name or two from your Contacts, it's easy to do. With Contacts open, tap the contact you want to delete.

2. On the information page (refer to **Figure 5-4**), tap the Edit button.

3. On the Info page that displays, drag your finger upward to scroll down and then tap the Delete Contact button at the bottom (see **Figure 5-23**).

4. The confirming dialog shown in **Figure 5-24** appears; tap the Delete Contact button to confirm the deletion.

Figure 5-23

Tap this button

Figure 5-24

 During this process, if you change your mind before you tap Delete, tap the Cancel button in Step 4. But be careful: After you tap Delete, there's no going back!

Using Handy Utilities

*i*Phone has a Utilities folder which you'll find on the second Home screen. Tucked away here are three little apps that can be very useful indeed.

In this chapter I help you explore Calculator to keep your numbers in line, Compass to help you find your way, and Voice Memos so you can record your best ideas for posterity.

Use the Calculator

1. This one won't be rocket science. The Calculator app works just about like every calculator app you've ever used. Go to the second Home screen and locate the Utilities folder.

2. Tap the Utilities folder to display the contents shown in **Figure 6-1**.

Figure 6-1

3. Tap the Calculator icon. The Calculator app appears.

4. Tap a few numbers (see **Figure 6-2**) and then use any of these functions and additional numbers to perform calculations:

⟼ **+, −, x, and ÷:** These familiar buttons add, subtract, multiply, and divide the number you've entered.

⟼ **mc, m+, m-, mr:** These are memory functions. You can, in order, clear memory, add a calculation into memory, deduct a calculation from memory, or retrieve the memory results.

⟼ **+/−:** If the calculator is displaying a negative result, tap this to change it to a positive result, and vice versa.

⟼ **AC/C:** This is the clear button; its name will change depending on whether you've entered one item or several (AC clears all; C clears just the last entry).

➠ =: This button produces the result of whatever calculation you've entered.

Figure 6-2

 If you have a scientific nature you'll be delighted to see that if you turn your phone to a landscape orientation you get additional features that turn the basic calculator into a scientific calculator so you can play with calculations such as cosigns and tangents.

Find Your Way with Compass

1. Compass is a handy way to find your way out of the woods, assuming that you get 3G reception out there. Tap the Utilities folder and then tap the Compass app. The first time you do this, a message appears, asking if iPhone can use your current location to provide information. Tap OK.

2. The Compass app appears (see **Figure 6-3**). Move around with your iPhone, and the compass indicates your current orientation in the world.

3. Tap the Information icon in the bottom-right corner and, from the choices that appear as shown in **Figure 6-4,** choose True North or Magnetic North.

Maps icon   Information icon

Figure 6-3

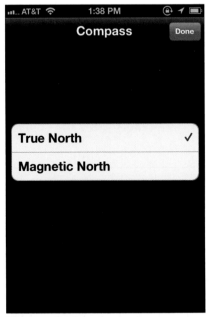

Figure 6-4

4. To go to your current location in the Maps app with the compass feature of that app activated (see **Figure 6-5**), tap the Maps icon in the bottom-left corner of the Compass screen (refer to **Figure 6-3**).

 True North refers to the direction you follow from where you are to get to the North Pole; Magnetic North is correlated relative to the Earth's magnetic field. True North is the more accurate measurement because of the tilt of the Earth.

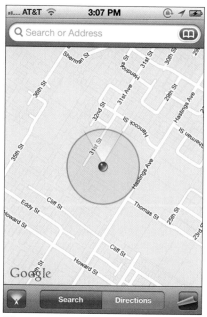

Figure 6-5

Record Voice Memos

1. Voice Memos is perhaps the most robust of the three apps covered in this chapter. The app allows you to record memos, edit memos by trimming them down, share them via e-mail or instant message with iMessage, and label recordings so you find them again. In the Utilities folder, tap the Voice Memos icon.

2. In the Voice Memo app (see **Figure 6-6**), tap the red Record button to record a memo. (This button changes to a red Pause button when you are recording.)

3. The top of the screen tells you that you're in recording mode (see **Figure 6-7**). While recording, you can tap the red Pause button to pause the recording or tap the black Stop button to stop recording.

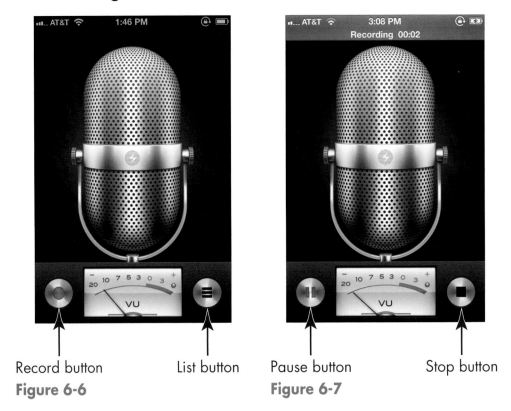

Record button   List button   Pause button   Stop button

Figure 6-6   **Figure 6-7**

4. When you're done recording, tap the List button (refer to **Figure 6-6**) and a list of recorded memos appears (see **Figure 6-8**).

 You can delete a memo you no longer need or want from the list of recorded memos (refer to **Figure 6-8**). To do so, tap on the memo in the list you want to delete and then tap the Delete button.

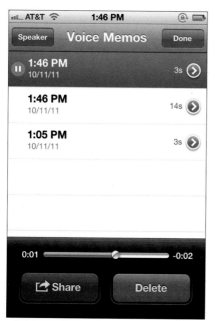

Figure 6-8

Trim a Voice Memo

1. Perhaps you repeated yourself at the beginning of your memo. If you want to cut part of a recorded memo out, you can trim it. With the list of recordings displayed (refer to **Figure 6-8**), tap the arrow on the right of any recording. The recording details as shown in **Figure 6-9** display.

2. Tap Trim Memo.

3. In the Trim Voice Memo screen shown in **Figure 6-10**, tap on the right or left of the memo bar and drag inward to trim a portion of the recording.

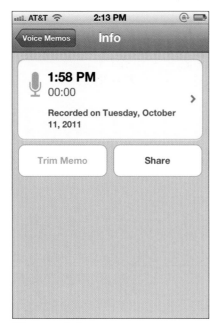

Figure 6-9

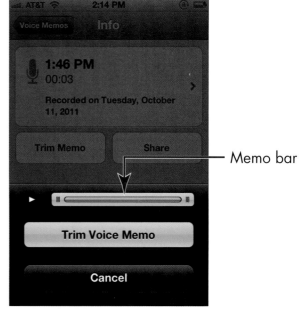

Memo bar

Figure 6-10

4. Tap the Trim Voice Memo button to complete the trim action. You return to the memo details dialog shown in **Figure 6-9.** Tap the Voice Memos button to return to the list of memos.

 If you begin to trim a memo and change your mind, tap the Cancel button.

Label a Voice Memo

1. It may help you find a voice memo if you label it. For example, you can label a memo as an idea, a meeting memo, or an interview. With the list of memos displayed (refer to **Figure 6-8**), tap the arrow to the right of a memo to display its details.

2. Tap the arrow on the memo information field to display the Label options shown in **Figure 6-11.**

3. Tap a label (or tap the Custom option and create your own label) and then tap the Info button to return to the memo information screen, where you see the memo now named with the label you just gave it (see **Figure 6-12**).

Figure 6-11

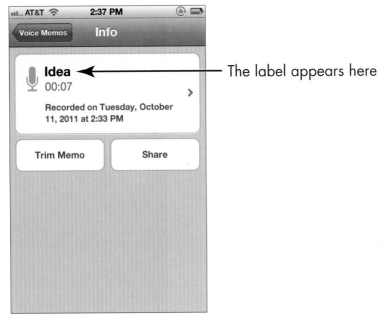

The label appears here

Figure 6-12

Share a Voice Memo

1. Tap a memo in the list of memos (refer to **Figure 6-8**) to select it.

2. Tap the Share button shown in **Figure 6-13**.

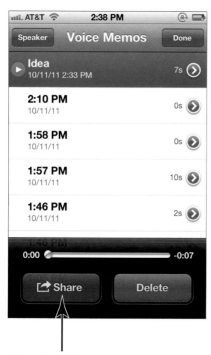

Tap this button

Figure 6-13

3. In the menu shown in **Figure 6-14,** tap Email to display an e-mail form or Message to display a Messages form to send an instant message with the voice memo attached (see **Figure 6-15**).

Figure 6-14

Figure 6-15

4. Fill in the recipient's information, a subject if you're sending an e-mail, and a message, and tap Send. The voice memo and your message go on their way.

Taking Advantage of Accessibility Features

*i*Phone users are all different; some face visual, dexterity, or hearing challenges. If you're one of these folks, you'll be glad to hear that iPhone offers some handy accessibility features.

To make your screen easier to read, you can adjust the brightness or change wallpaper. The Zoom feature lets you enlarge the screen even more than the standard unpinch gesture does. The black-and-white screen option even offers a black background with white lettering that some people find easier to use when reading text. You can also set up the VoiceOver feature to read onscreen elements out loud.

If hearing is your challenge, you can do the obvious and adjust the system volume. The iPhone also has a setting for mono audio that's useful when you're wearing headphones. Rather than breaking up sounds in a stereo effect, this setting puts all the sound in each ear. If you have more trouble hearing in one ear than in the other, this option can help make sounds clearer. iPhone uses the Speak

Auto-text feature to tell you whenever text you enter in any iPhone application is autocorrected or capitalized. Finally, you can set a larger text size for your iPhone to bring text into better focus and turn on the AssistiveTouch control panel to help you with making gestures and choices if you struggle with the touchscreen interface.

Set Brightness

1. Especially when using iPhone as an e-reader, you may find that a slightly less-bright screen reduces strain on your eyes. To begin, tap the Settings icon on the Home screen.

2. In the Settings dialog,, tap Brightness.

3. To control brightness manually, tap the Auto-Brightness On/Off button (see **Figure 7-1**) to turn off this feature.

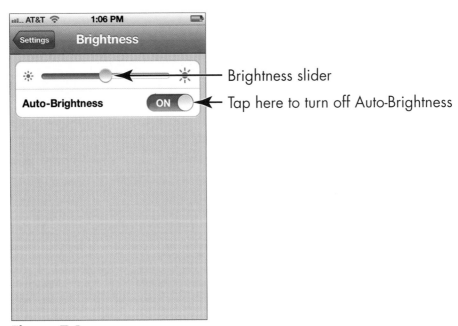

Figure 7-1

4. Tap and drag the Brightness slider (refer to **Figure 7-1**) to the right to make the screen brighter, or to the left to make it dimmer.

5. Tap the Home button to close the Settings dialog.

 If glare from the screen is a problem for you, consider getting a *screen protector*. This thin film not only protects your screen from damage but can also reduce glare.

 In the iBooks e-reader app, you can set a sepia tone for the page, which might be easier on your eyes. See Chapter 14 for more about using iBooks.

Change the Wallpaper

1. The picture of water drops — the default iPhone background image — may be pretty, but it may not be the one that works best for you. Choosing different wallpaper may help you to see all the icons on your Home screen. Start by tapping the Settings icon on the Home screen.

2. In the Settings dialog, tap Wallpaper.

3. In the Wallpaper settings that appear, tap the arrow to the right of the iPhone images and then tap Wallpaper.

4. The wallpaper options shown in **Figure 7-2** appear. Tap one to select it.

5. In the preview that appears (see **Figure 7-3**), tap Set and in the menu that appears tap either Set Lock Screen (the screen that appears when you lock the iPhone by tapping the power button), Set Home Screen, or Set Both.

Figure 7-2

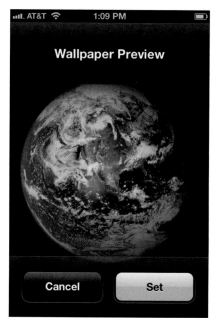

Figure 7-3

6. Tap the Home button to return to your Home screen with the new wallpaper set as the background.

Turn On Zoom

1. The Zoom feature enlarges the contents displayed on the iPhone screen when you double-tap the screen with three fingers. Tap the Settings icon on the Home screen and then tap General. In General settings, drag your finger up to scroll down the screen and tap Accessibility. The Accessibility pane, shown in **Figure 7-4,** appears.

2. Tap Zoom (refer to **Figure 7-4**).

3. In the Zoom pane shown in **Figure 7-5,** tap the Zoom On/Off button to turn on the feature.

Tap this option

Figure 7-4

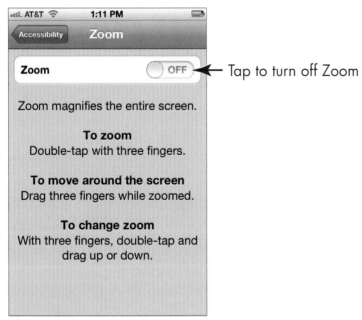

Tap to turn off Zoom

Figure 7-5

4. Tap the Home button to close Settings. At this point screens will be enlarged (zoomed). You need to swipe with three fingers to scroll down the Home screen and locate the icons in the Dock along the bottom.

5. Tap Safari and go to a website (www.wiley.com, for example) and double-tap the screen using three fingers; it enlarges.

6. Press three fingers on the screen and drag to move it around.

7. Double-tap with three fingers again to go back to regular magnification.

8. Tap the Home button to close Settings.

 The Zoom feature works almost everywhere in iPhone: in Photos, on web pages, on your Home screens, in your Mail, in Music, and in Videos — give it a try!

Turn On White on Black

1. The White on Black accessibility setting reverses colors on your screen so that backgrounds are black and text is white. To turn on this feature, tap the Settings icon on the Home screen.

2. Tap General and then scroll down and tap Accessibility.

3. In the Accessibility dialog, tap the White on Black On/Off button to turn on this feature (see **Figure 7-6**).

Tap to turn on White on Black

Figure 7-6

4. The colors on the screen reverse. Tap the Home button to leave Settings.

The White on Black feature works well in some places and not so well in others. For example, in the Photos application, pictures appear almost as photo negatives. Your Home screen image will likewise look a bit strange. And don't even think of playing a video with this feature turned on! However, if you need help reading text, White on Black can be useful in several applications.

Set Up VoiceOver

1. VoiceOver reads the names of screen elements and settings to you, but it also changes the way you provide input to the iPhone. In Notes, for example, you can have VoiceOver read the name of the Notes buttons to you and, when you enter notes, it will read words or characters you've entered. It can also tell you whether features such as Auto-Correction are on. To turn on this feature, tap the Settings icon on the Home screen. Tap General and then scroll down and tap Accessibility.

2. In the Accessibility pane shown in **Figure 7-7,** tap the VoiceOver button.

3. In the VoiceOver pane shown in **Figure 7-8,** tap the VoiceOver On/Off button at the top to turn on this feature. You see a dialog noting that turning on VoiceOver changes gestures that you use to interact with the iPhone. Tap OK once to select the button, and then tap OK twice to proceed.

With VoiceOver on, you must first single-tap to select an item such as a button, which causes VoiceOver to read the name of the button to you. Then you double-tap the button to activate its function. (VoiceOver reminds you about this if you turn on Speak Hints, which is helpful when you first use VoiceOver, but it soon becomes annoying.)

Figure 7-7

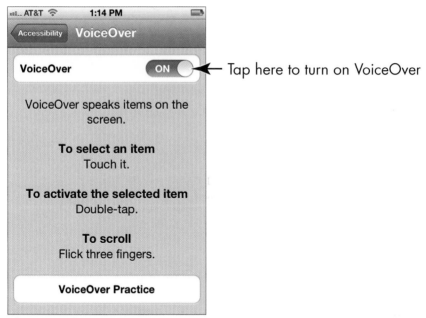

Tap here to turn on VoiceOver

Figure 7-8

4. Tap the VoiceOver Practice button to select it, and then double-tap the button to open VoiceOver Practice. (This is the new method of tapping that VoiceOver activates.) Practice using gestures such as pinching or flicking left, and VoiceOver tells you what action each gesture initiates.

5. Tap the Done button and then double-tap it to return to the VoiceOver dialog. Tap the Speak Hints field, and VoiceOver speaks the name of the item. Double-tap the slider to turn off Speak Hints.

6. If you want VoiceOver to read words or characters to you (for example, in the Notes app), double-tap Typing Feedback.

7. In the Typing Feedback dialog, tap to select the option you prefer. The Words option causes VoiceOver to read words to you, but not characters, such as "dollar sign" ($). The Characters and Words option causes VoiceOver to read both.

8. Tap the Home button to return to the Home screen. Read the next task to find out how to navigate your iPhone after you've turned on VoiceOver.

 You can change the language that VoiceOver speaks. In General settings, choose International and then Language and select another language. This action, however, also changes the language used for labels on Home icons and various settings and fields in iPhone.

 You can use the Set Triple-Click Home setting to help you more quickly turn the VoiceOver, Zoom, or White on Black features on and off. In the Accessibility dialog, tap Triple-Click Home. In the dialog that appears, choose what you want a triple-click of the Home button to do: toggle VoiceOver on or off; toggle White on Black on or off; toggle the Zoom feature on or off; toggle on the AssistiveTouch Control panel; or display a menu of options using the Ask choice. Now a triple-click with a single finger on the Home button provides you with the option you selected wherever you go in iPhone.

Use VoiceOver

After VoiceOver is turned on, you need to figure out how to use it. I won't kid you; using it is awkward at first, but you'll get the hang of it! Here are the main onscreen gestures you should know how to use:

⟶ **Tap an item to select it.** VoiceOver then speaks its name.

⟶ **Double-tap the selected item.** This action activates the item.

⟶ **Flick three fingers.** It takes three fingers to scroll around a page with VoiceOver turned on.

Table 7-1 provides additional gestures to help you use VoiceOver. I suggest that, if you want to use this feature often, you read the VoiceOver section of the iPhone online *User Guide*, which goes into a great deal of detail about the ins and outs of using VoiceOver. You'll find the User Guide in the Bookmarks section of the Safari browser.

 If tapping with two or three fingers seems difficult for you, try tapping with one finger from one hand and one or two from the other. When double- or triple-tapping, you have to perform these gestures as quickly as you can for them to work.

 Check out some new settings in iOS 5 for VoiceOver, including a choice for Braille, language choices, the ability to navigate images, and a setting to have iPhone speak notifications.

Table 7-1	VoiceOver Gestures
Gesture	**Effect**
Flick right or left.	Select the next or preceding item.
Tap with two fingers.	Stop speaking the current item.
Flick two fingers up.	Read everything from the top of the screen.
Flick two fingers down.	Read everything from the current position.
Flick three fingers up or down.	Scroll one page at a time.
Flick three fingers right or left.	Go to the next or preceding page.
Tap three fingers.	Speak the scroll status (for example, line 20 of 100).
Flick four fingers up or down.	Go to the first or last element on a page.
Flick four fingers right or left.	Go to the next or preceding section (as on a web page).

Adjust the Volume

1. Though individual apps such as Music and Video have
their own volume settings, you can set your iPhone system volume as well to help you better hear what's going
on. Volume settings in apps are based on a percentage of
the system volume setting. Tap the Settings icon on the
Home screen.

2. Tap Sounds.

3. In the Sounds pane that appears (see **Figure** 7-9), tap
and drag the slider to the right to increase the volume, or
to the left to lower it.

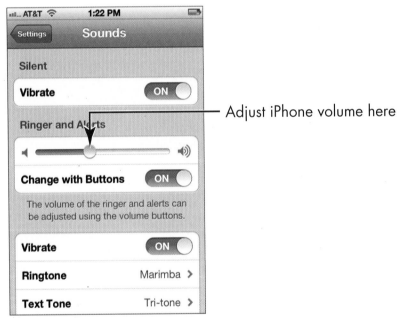

Adjust iPhone volume here

Figure 7-9

4. Tap the Home button to close Settings.

In the Sounds pane, you can turn on or off the sounds that iPhone makes when certain events occur (such as receiving new mail or Calendar alerts). These sounds are turned on by default. Conversely, if you need the audio clue and this feature has been turned off, just tap the On/Off button for the item to turn it back on.

 Even those of us with perfect hearing sometimes have trouble hearing a phone ring, especially in busy public places. Consider using the Vibrate setting in the Sounds pane to have your phone vibrate when a call is coming in.

Use Mono Audio

1. Using the stereo effect in headphones or a headset breaks up sounds so that you hear a portion in one ear and a portion in the other ear, to simulate the way your ears process sounds. However, if you're hard of hearing or deaf in one ear, you're hearing only a portion of the sound in your hearing ear, which can be frustrating. If you have such hearing challenges and want to use iPhone with a headset connected, you should turn on Mono Audio. When it's turned on, you can set up iPhone to play all sounds in each ear. Tap the Settings icon on the Home screen.

2. Tap General and then scroll down and tap Accessibility.

3. In the Accessibility pane, scroll down and tap the Mono Audio On/Off button (see **Figure 7-10**) to turn on the feature.

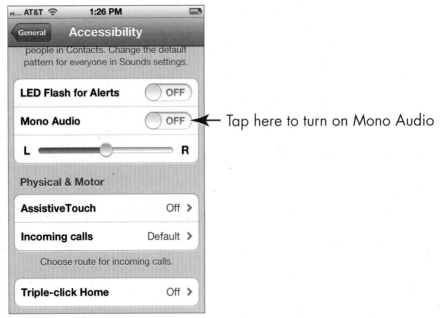

Figure 7-10

4. Tap and drag the slider to L for sending sound to only your left ear or R for right ear.

5. Tap the Home button to close Settings.

 If you have hearing challenges, another good feature that iPhone provides is support for closed-captioning. In the video player, you can use the closed-captioning feature to provide onscreen text representing dialogue and actions in a movie (if it supports closed-captioning) as it plays. For more about playing videos, see Chapter 17.

Have iPhone Speak Auto-text

1. The Speak Auto-text feature speaks autocorrections and autocapitalizations (two features that you can turn on using Keyboard settings). When you enter text in an app such as Notes or Mail, the app then makes either type of change, while Speak Auto-text lets you know what change was made. To turn on Speak Auto-text, tap the Settings icon on the Home screen.

2. Tap General and then scroll down and tap Accessibility.

3. In the Accessibility pane shown in **Figure 7-11,** tap the Speak Auto-text On/Off button to turn on the feature.

Tap here to turn on Speak Auto-text

Figure 7-11

4. Tap the Home button to leave Settings.

Why would you want iPhone to tell you whenever an autocorrection has been made? If you have vision challenges and you know that you typed *ain't* when writing dialogue for a character in your novel, but iPhone corrected it to *isn't,* you would want to know. Similarly, if you type the poet's name *e.e. Cummings* and autocapitalization corrects it (incorrectly), you need to know immediately so that you can change it back again!

Turn On Large Text

1. If having larger text in the Contacts, Mail, and Notes apps would be helpful to you, you can turn on the Large Text feature and choose the text size that works best for you. To turn on Large Text, tap the Settings icon on the Home screen.

2. Tap General and then scroll down and tap Accessibility.

3. In the Accessibility pane, tap the Large Text button to turn on the feature.

4. In the list of text sizes shown in **Figure 7-12,** tap the one you prefer.

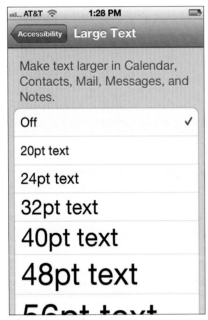

Figure 7-12

5. Tap the Home button to close the Settings dialog.

Turn On and Work with AssistiveTouch

1. The AssistiveTouch Control panel helps those who have challenges working with buttons to provide input to iPhone using the touchscreen. To turn on AssistiveTouch, tap Settings on the Home screen and then tap General and Accessibility.

2. In the Accessibility pane, scroll down and tap AssistiveTouch. In the pane that appears, tap the On/Off button for AssistiveTouch to turn it on (see **Figure** 7-13). Note the gray square (called the AssistiveTouch Control panel) that appears on the right side of the pane. This square now appears on the screen in whatever apps you display on your iPhone.

3. Tap the AssistiveTouch Control panel to display options, as shown in **Figure** 7-14.

— Tap to turn on AssistiveTouch

— AssistiveTouch Control panel

Figure 7-13

Figure 7-14

4. You can tap Gestures, Favorites, or Device on the panel to see additional choices, or tap Home to go directly to the Home screen. Once you've chosen an option, tapping the Back arrow takes you back to the main panel.

Table 7-2 shows the major options available in the AssistiveTouch Control panel and their purpose:

Table 7-2	AssistiveTouch Controls
Control	*Purpose*
Gestures	Choose the number of fingers to use for gestures on the touchscreen.
Favorites	Displays a set of gestures with only the Pinch gesture preset; you can tap any of the other blank squares to add your own favorite gestures.
Device	You can rotate the screen, lock the screen, turn volume up or down, mute or unmute sound, or shake iPhone to undo an action using the presets in this option.
Home	Sends you to the Home screen.

Talk to Your iPhone with Siri

Though it has a faster processor and higher-quality camera, the real reason to buy iPhone 4S is Siri, a personal assistant feature that responds to the commands you speak to your phone. With Siri, you can ask for nearby restaurants, and a list appears. You can dictate your e-mail messages rather than typing them. Calling your mother is as simple as saying, "Call Mom." Want to know the capital of Rhode Island? Just ask. Siri checks a built-in database to answer questions ranging from the result of a mathematical calculation to the size of Jupiter.

If you have an iPhone 4 or 3GS, you don't have Siri but you do have a Voice Control feature you can explore by holding down the Home button for about 2 seconds until you hear a beep. This feature is more limited, and you have to state commands in a very specific way such as "Call Joe" or "Play Music." Note that, with an iPhone 4S, when you activate Siri you override Voice Control. Which is just as well, because Siri leaves Voice Control in the dust.

If you don't have an iPhone 4S, you can skip this chapter and dream of the day when you can upgrade (or jailbreak your iPhone 4 and port Siri to it). But if you have your iPhone 4S in hand, you're about to discover an absolutely awesome feature.

Activate Siri

When you first go through the process of registering your phone, making settings for your location, using iCloud, and so on, at one point you will see the screen in **Figure 8-1**. To activate Siri at this point, just tap Use Siri. As you begin to use your phone, iPhone reminds you about using Siri by displaying the message shown in **Figure 8-2**.

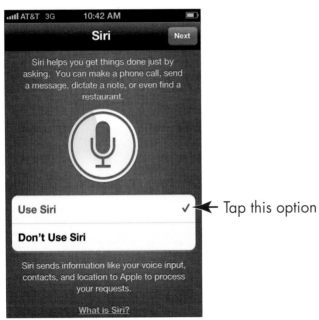

← Tap this option

Figure 8-1

Figure 8-2

If you didn't activate Siri during the registration process, you can use Settings to turn Siri on by following these steps:

1. Tap the Settings icon on the Home screen.

2. Tap General, and then tap Siri.

3. In the dialog in **Figure 8-3,** tap the On/Off button to turn Siri on.

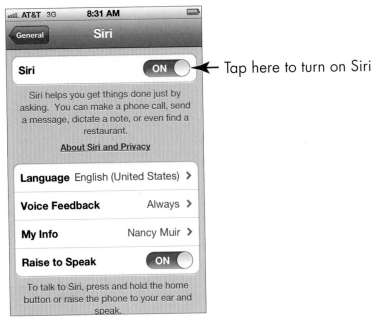

Tap here to turn on Siri

Figure 8-3

4. If you wish to change the language Siri uses, tap Language and choose a different language in the list that appears.

If you only want Siri to verbally respond to your requests when the handset isn't in your hands, Tap Voice Feedback and choose Handsfree Only. Here's how this setting works and why you might want to use it: In general if you're holding your iPhone you can read responses on the screen so you might choose not to have your phone talk to you out loud. In addition, if you are, say, puttering with an electronics project and want to speak requests for mathematical calculations and hear the answers rather than have to read them, Handsfree is a useful setting.

Siri is only available on iPhone 4S with Internet access and cellular data charges could apply. In addition, Apple warns that available features may vary by area.

Understand All that Siri Can Do

Siri allows you to interact with many apps on your iPhone by voice. You can pose questions or ask to do something like make a call or add an appointment to your calendar, for example. Siri can also search the Internet or use an informational service called Wolfram Alpha to provide information on just about any topic.

Siri knows what app you're using, though you don't have to be in an app to make a request involving another app. However, if you are in the Messages app, you can make a statement like, "Tell Susan I'll be late.", and Siri knows that you want to send a message.

Siri can also take dictation. When you have Siri activated, the onscreen keyboard contains a microphone key you can tap to begin dictation, and tap to end dictation. This works in any app that uses the onscreen keyboard.

Siri requires no preset structure for your questions; you can phrase things in several ways. For example, you might say, "Where am I?" to see a map of your current location, or you could say, "What is my current location?" or "What address is this?" and get the same results.

Siri responds to you both verbally and with text information, in a form as with e-mail (see **Figure 8-4**), or in a graphic display for some items such as maps. When a result appears, you can tap it to make a choice or open a related app.

Siri works with Phone, Music, Messages, Reminders, Calendar, Maps, Mail, Weather, Stocks, Clock, Contacts, Notes, Safari, and the Wolfram Alpha online service. In the following tasks, I provide a quick guide to some of the most useful ways you can use Siri.

Note that no matter what kind of action you wish to perform, first press and hold the Home button until Siri opens (see **Figure 8-5**). Remember this only works with iPhone 4S; for earlier iPhones this would activate Voice Control. The rest of this chapter assumes you're working with a 4S.

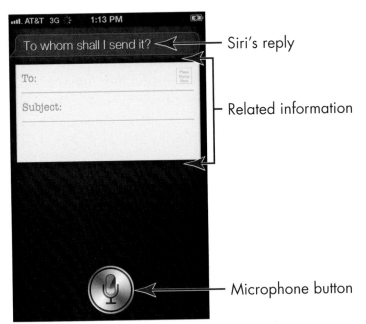

Siri's reply

Related information

Microphone button

Figure 8-4

Figure 8-5

Call Contacts

First make sure the people you call are entered in your Contacts app and include their phone number in their records. If you want to call somebody by stating your relationship to her, such as "Call sister.", be sure to enter that relationship in the related field in her contact record and make sure that the settings for Siri (refer to **Figure 8-3**) include your contact name in the My Info field. (See Chapter 5 for more about creating contact records.)

1. Press and hold the Home button until Siri appears.

2. Speak a command such as "Call Harold Smith." or "Call Mom." If you want to make a FaceTime call, you can say "FaceTime Mom."

3. If there are two contacts who might match a spoken name, Siri responds with a list of possible matches (see **Figure 8-6**). Tap one in the list or state the correct contact's name to proceed.

Figure 8-6

4. The call is placed. To end it before it completes, you can tap the Home button and then tap End.

 To cancel any spoken request, you have three options: You can say, "Cancel.", tap the Microphone button on the Siri screen, or tap the Home button.

Create Reminders and Alerts

1. You can also use Siri with the Reminders app. To create a reminder, press and hold the Home key and then speak a command, such as "Remind me to call Dad on Thursday at 10 a.m." or "Wake me up tomorrow at 7 a.m."

2. A preview of the reminder is displayed (see **Figure 8-7**), and Siri asks you if it should create the reminder. Tap or say Confirm to create it or Cancel.

Figure 8-7

3. If you want a reminder ahead of the event you created, activate Siri and speak a command, such as "Remind me tonight about the play on Thursday at 8 p.m." A second reminder is created, which you can confirm or cancel.

Add Tasks to Your Calendar

1. You can also set up events on your Calendar using Siri. Press and hold the Home button and then speak a phrase, such as "Set up meeting at 10 a.m. on October 12th."

2. A sample calendar entry appears, and Siri asks if you want to confirm it.

3. If there's a conflict with the appointment, Siri tells you that there's already an appointment at that time (see **Figure 8-8**) and asks if you still want to set up the new appointment. You can say Yes or Cancel at that point or tap the Yes or Cancel button.

Figure 8-8

Play Music

1. Press and hold the Home button until Siri appears.

2. To play music, speak a command, such as "Play music." or "Play 'As Time Goes By'." to play a specific song or album.

3. When music is playing, use commands, such as "Pause music.", "Next track.", or "Stop music." to control playback.

 One of the beauties of Siri is that you don't have to follow a specific command format as you do with Voice Control. You could say "Play the next track" or "Next track" or "Jump to the next track on this album" and Siri will get your meaning. If you're stuck with Voice Control, check the *iPhone User Guide* for the correct syntax for all commands.

Get Directions

You can use the Maps app and Siri to find your current location, find nearby businesses such as restaurants or a bank, or get a map of another location.

Here are some of the commands you can try to get directions or a list of nearby businesses:

➡ **"Where am I?"**

Displays a map of your current location.

➡ **"Where is Apache Junction, Arizona?"**

Displays a map of that city.

⟶ "Find restaurants."

Displays a list of restaurants near your current location as in **Figure** 8-9; tap one to display a map of its location.

Figure 8-9

⟶ "Find Bank of America."

Displays a map with the location of that business (or in some cases several nearby locations, such as a bank branch and all ATMs) indicated.

 Once a location is displayed in a map, tap the information button on the location's label to view its address, phone number, and website address, if available.

Ask for Facts

Wolfram Alpha is a self-professed online computational knowledge engine. That means it's more than a search engine because it provides specific information about a search term rather than multiple search results. If you want facts without having to spend time browsing websites to find those facts, Wolfram Alpha is a very good resource.

Siri uses Wolfram Alpha to look up facts in response to questions, such as "What is the capital of Kansas?", "What is the square root of 2003?", or "How large is Mars?" Just press and hold the Home button and ask your question; Siri consults Wolfram Alpha and returns a set of relevant facts.

You can also get information about the weather, stocks, or the time. Just say a phrase like one of these to get what you need:

➡ **"What is the weather?"**

This shows the weather report for your current location. If you want weather in another location, just specify the location in your question.

➡ **"What is the price of Apple stock?"**

Siri tells you the current price of the stock or the price of the stock when the stock market last closed.

➡ **"What time is it?"**

Siri tells you the time and displays a clock (see **Figure 8-10**).

Figure 8-10

Search the Web

While Siri can use Wolfram Alpha to respond to specific requests such as "Who is the Queen of England?," more general requests for information will cause Siri to search the web.

For example, if you speak a phrase, such as "Find a website about birds." or "Find information about the World Series.", Siri can respond in a couple of ways. The app can simply display a list of search results using the default search engine specified in your settings for Safari or suggesting, "If you like, I can search the web for such-and-such." In the first instance, just tap a result to go to that website. In the second instance, you can confirm that you want to search the web or cancel.

Send E-mail or Messages

You can create an e-mail or an instant message using Siri. For example, if you say, "E-mail Jack Wilkes.", a form opens already addressed to that contact. Siri asks you what to say; speak your message and then say, "Send." to speed your message on its way.

Siri also works with the iMessage feature. If you have the Messages app open and you say, "Tell Sarah I'll call soon.", Siri creates a message for you to approve and send.

 At this point in time, Siri can't tweet. But if you try, she apologizes for the inconvenience!

Use Dictation

1. With Siri active, a microphone key appears on the onscreen keyboard (see **Figure 8-11**) so you can use a dictation feature to speak text rather than type it. Go to any app where you enter text, such as Notes or Mail, and tap in the document or form. The onscreen keyboard appears.

Microphone key

Figure 8-11

2. Tap the microphone key on the keyboard and speak your text.

3. To end the dictation, tap the microphone key again.

 When you finish speaking text, you can use the keyboard to make edits to the text Siri entered, although as voice recognition programs go, Siri is pretty darn accurate. If a word sports a blue underline, which means there may be an error, you can tap to select and make edits to it.

Get Helpful Tips

I know you're going to have a wonderful time learning the ins and outs of Siri, but before I close this chapter, here are some tips to get you going:

➡ **Activating Raise to Speak:** Go to Settings, General, and under the Siri setting use the Raise to Speak setting to activate Siri when you put your phone to your ear rather than pressing and holding the Home button.

➡ **If Siri doesn't understand you:** When you speak a command and Siri displays what it thought you said, if it misses the mark, you have a few options. To correct a request you've made, you can tap the bubble containing the command Siri heard and edit it by typing or tapping the microphone key on the onscreen keyboard and dictating the correct information. If a word is underlined in blue, it's a possible error. Tap the word and then tap an alternative that Siri suggests. You can also simply speak to Siri and say something like, "I meant Sri Lanka." or "No, send it to Sally." If even corrections aren't working you may need to restart your phone to reset the software.

➡ **Headsets and earphones:** If you're using iPhone earphones or a Bluetooth headset to activate Siri, instead of pressing the Home button press and hold the center button (the little button on the headset that starts and stops a call).

➡ **Using Find My Friends:** There is a free app you can download from the App Store called Find My Friends that allows you to ask Siri to locate your friends geographically.

➡ **Getting help:** To get help with Siri features, just press and hold the Home button and ask Siri, "What can you do?"

Getting Social with FaceTime, Twitter, and iMessage

FaceTime is an excellent video-calling app that's been available on iPhone 4 since its release in mid-2010. The app lets you call people who have FaceTime on their devices — whether iPhone 4 or 4S, iPad 2, fourth-generation iPod touch, or Mac (running Mac OS X 10.6.6 or later) — using either a phone number (iPhone) or an e-mail address (Mac, iPod touch, or iPad 2). You and your friend or family member can see each other as you talk, which makes for a much more personal calling experience.

Twitter is a social networking service referred to as a *microblog,* because it involves only short posted messages. Twitter has been incorporated into iOS 5 in a way that allows you to tweet people from within the Safari, Photos, Camera, YouTube, and Maps apps. You can also download the free Twitter app and use it to post tweets whenever you like.

Finally, iMessage is a feature available through the preinstalled Messages app for instant messaging (IM). IM involves sending a text message to somebody's iPhone (using her phone number) or iPod touch or iPad (using her e-mail address) to carry on an instant conversation.

In this chapter, I introduce you to FaceTime, Twitter, and iMessage and review their simple controls. In no time, you'll be socializing with all and sundry.

Get an Overview of FaceTime

FaceTime lets you call other folks who have a FaceTime device (iPhone 4 or 4S, iPad 2, a fourth-generation iPod touch, or a compatible Mac computer) and chat while sharing video images of each other. This feature uses the Camera app and is useful for seniors who want to keep up with distant family members and friends and see (as well as hear) the latest-and-greatest news.

At this point, you can make and receive calls with FaceTime using a phone number (on iPhone 4 or 4S) or an e-mail account (iPad 2, iPod touch, or Mac) and show the person on the other end what's going on around you. Just remember that you can't adjust audio volume from within the app or record a video call. Nevertheless, on the positive side, even though its features are limited, this app is straightforward to use.

You can use your Apple ID and e-mail address to access FaceTime, so it works pretty much right away. See Chapter 3 for more about getting an Apple ID.

 If you're having trouble using FaceTime, make sure the FaceTime feature is turned on. That's quick to do: Tap Settings on the Home screen, tap FaceTime, and then tap the On/Off button to turn it on, if necessary. You can also select the e-mail account that can be used to make phone calls to you on this Settings page.

Make a FaceTime Call

1. If you know that the person you're calling has FaceTime on an iPhone, an iPad 2, or a Mac, add that person to your iPhone Contacts (see Chapter 5 for how to do this) if he or she isn't already in there.

2. Tap the Contacts app icon on the second Home screen.

3. Scroll to locate a contact and tap the contact's name to display their information (see **Figure** 9-1).

4. Tap the FaceTime button. If the contact has both an e-mail address and phone number, a dialog displays them both. Tap on the one you want to use. You've just placed a FaceTime call! (Note if the contact has only a phone or e-mail recorded, when you tap the FaceTime button the call goes through immediately.)

 You have to use the appropriate method for placing a FaceTime call, depending on the kind of device the person you're calling has. If you're calling an iPhone 4 or 4S user, you should use a phone number the first time you call and thereafter you can use the phone number or e-mail address; if you're calling an iPad 2, an iPod touch, or a FaceTime for Mac user, you have to make the call using that person's e-mail address.

 When you call somebody using an e-mail address, the person must be signed in to his Apple ID account and have verified that the address can be used for FaceTime calls. iPhone 4 and 4S, iPad 2, and iPod touch (fourth generation) users can make this setting by tapping Settings and then FaceTime; FaceTime for Mac users make this setting by selecting FaceTime⇨Preferences.

5. When the person accepts the call, you see a large screen that displays the recipient's image and a small screen containing your image superimposed in one corner (see **Figure** 9-2).

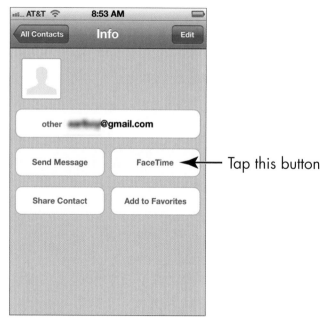

Tap this button

Figure 9-1

Figure 9-2

To view recent calls, tap the Phone app icon and then tap the Recents button. Tap a recent call and iPhone displays that person's information. You can tap the contact to call the person back.

At the time of this writing, you can use FaceTime only over a Wi-Fi network, not over 3G, which limits the places from which you can make or receive video calls to your home wireless network or a public hotspot.

Accept or End a FaceTime Call

1. If you're on the receiving end of a FaceTime call, accepting the call is about as easy as it gets. When the call comes in, tap the Accept button to take the call or tap the Decline button to reject it (see **Figure 9-3**).

Figure 9-3

2. Chat away with your friend, swapping video images. To end the call, tap the End button (see **Figure 9-4**).

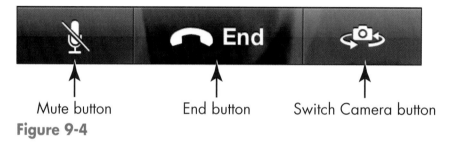

| Mute button | End button | Switch Camera button |

Figure 9-4

To mute sound during a call, tap the Mute button (refer to **Figure** 9-4). Tap the button again to unmute your iPhone.

Switch Views

1. When you're on a FaceTime call, you might want to use iPhone's built-in camera to show the person you're talking to what's going on around you. Tap the Switch Camera button (refer to **Figure** 9-4) to switch from the front-facing camera that's displaying your image to the back-facing camera that captures whatever you're looking at (see **Figure** 9-5).

Figure 9-5

2. Tap the Switch Camera button again to switch back to the front camera displaying your image.

Experience Twitter on iPhone

Twitter is a social networking service for *microblogging,* which involves posting very short messages (limited to 140 characters) online so your friends can see what you're up to. You can go to www.twitter.com to sign up with the service; there is also a free Twitter app for iPhone that you can download and use to manage your Twitter account. Once you have an account, you can post tweets for all to see, have people follow your tweets, and follow the tweets that other people post.

With iOS 5, the ability to tweet has become integrated into several apps. You can post tweets using the Menu button within Safari, Photos, Camera, YouTube, and Maps. First, sign up for an account, and then download the free Twitter app to your iPhone. (See Chapter 13 for more about downloading apps.) Go to iPhone Settings and tap Twitter. Then add your account information.

Now when you're using Safari, Photos, Camera, YouTube, or Maps, you can choose Tweet from a menu. You'll see a Tweet form like that shown in **Figure 9-6.** Just write your message in the form and then tap Send.

 See Chapters 10, 16, or 17 for more about tweeting in the Safari, Photos, or YouTube apps.

Enter your message here...

then tap Send

Figure 9-6

Set Up an iMessage Account

1. iMessage is a feature available through the preinstalled
Messages app that allows you to send and receive instant
messages (IMs) to others using an Apple iOS device.
Instant messaging differs from e-mail or tweeting in an
important way. Where you might e-mail somebody and
wait days or weeks before that person responds, or you
might post a tweet that could sit there awhile before anybody views it, with instant messaging communication
happens immediately. You send an IM, and it appears on
somebody's iPhone, iPod touch, or iPad right away —
and assuming the person wants to participate in a live
conversation, the chat begins immediately, allowing a
back-and-forth dialog in real time. To set up iMessage,
tap Settings on the Home screen.

2. Tap Messages, and the settings shown in **Figure** 9-7 are displayed.

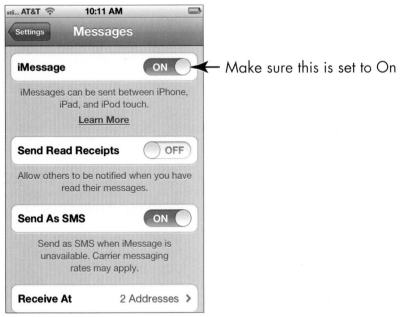

Make sure this is set to On

Figure 9-7

3. If iMessage isn't set to On (refer to **Figure 9-7**), tap the On/Off button to turn it on.

4. Check to be sure the e-mail account associated with your iPhone under the Receive At setting is correct. (This should be set up automatically based on your Apple ID.)

5. To allow a notice to be sent when you've read somebody's messages, tap the On/Off button for Send Read Receipts.

6. Tap the Home button to leave the settings.

To change the e-mail account used by iMessage, tap Receive At, tap the e-mail address, and then tap Remove This Email; then follow the preceding steps to add another e-mail account.

Use iMessage to Address, Create, and Send Messages

1. Now you're ready to use iMessage. From the Home screen, tap the Messages icon. Tap the New Message button in the top-right corner (see **Figure 9-8**) to begin a conversation.

2. In the form that appears (see **Figure 9-9**), you can address a message in a couple of ways:

- Begin to type an address in the To: field, and a list of matching contacts appears.

- Tap the plus icon on the right side of the address field, and the All Contacts list is displayed.

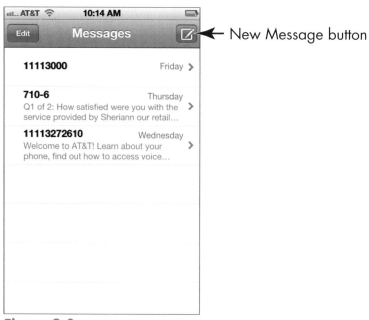

Figure 9-8

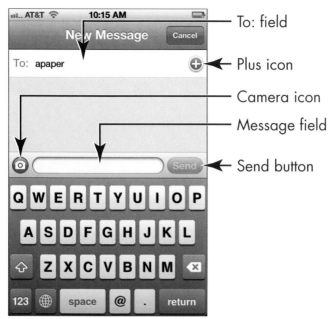

To: field

Plus icon

Camera icon

Message field

Send button

Figure 9-9

3. Tap a contact on the list you chose to display in the preceding step. If the contact has both an e-mail address and phone number stored, the Info dialog appears, allowing you to tap one or the other, which addresses the message.

4. To create a message, simply tap in the message field near the bottom of the screen (refer to **Figure** 9-9) and type your message.

5. To send the message, tap the Send button (refer to **Figure** 9-9). When your recipient(s) responds, you'll see the conversation displayed on the screen, as shown in **Figure 9-10.**

Edit button

The conversation is displayed here

Figure 9-10

 You can address a message to more than one person by simply choosing more recipients in Step 2 of the preceding list.

 If you want to include a photo or video with your message, tap the Camera icon to the left of the message field (refer to **Figure 9-9**). Tap Take Photo or Video or Choose Existing (depending on whether you want to create a new photo/video or send one you've already taken), and then tap Use. When you send your message, the photo or video will go along with your text.

Clear a Conversation

1. When you're done chatting, you might want to clear a conversation to remove the clutter before you start a new one. With Messages open and a conversation displayed, tap the Edit button (refer to **Figure 9-10**).

2. Tap the Clear All button (see **Figure** 9-11).

Clear All button

Figure 9-11

3. Tap Clear Conversation.

 You can also tap the Messages button to display all messages, tap the Edit button, and then tap the red Delete button next to any conversation you want to delete. Tap Done when you're finished.

 You can tap the Edit button for a conversation and then select all or part of the conversation to print, forward, or e-mail to somebody else.

Part III
Taking the Leap Online

The 5th Wave   By Rich Tennant

iPhone

"In fact it does come with a compass."

Browsing the Internet with Safari

Getting on the Internet with your iPhone is easy, using its Wi-Fi or 3G capabilities. After you're online, the built-in browser, *Safari*, is your ticket to a wide world of information, entertainment, education, and more. Safari will look familiar to you if you've used it on a PC or Mac device before, though the way you move around by using the iPhone touchscreen might be new to you. If you've never used Safari, this chapter takes you by the hand and shows you all its ins and outs.

In this chapter, you discover how to go online with your iPhone and navigate among web pages. Along the way, you see how to place a bookmark for a favorite site or place a web clip on your Home screen. You can also view your browsing history, save online images to your Photo Library, or e-mail or tweet a hotlink to a friend. You explore the Safari Reader and Safari Reading List features and learn how to keep yourself safer while online using private browsing. Finally, you review the simple steps involved in printing what you find online.

Get ready to . . .

Connect to the Internet

How you connect to the Internet depends on what connections are available:

➡ You can connect to the Internet via a Wi-Fi network. You can set up this type of network in your own home using your computer and some equipment from your Internet provider. You can also connect over public Wi-Fi networks, referred to as *hotspots*. You'll probably be surprised to discover how many hotspots your town or city has: Look for Internet cafés, coffee shops, hotels, libraries, and transportation centers such as airports or bus stations, for example. Many of these businesses display signs alerting you to their free Wi-Fi.

➡ You can also use the paid data network provided by AT&T, Sprint, or Verizon to connect using 3G from just about anywhere you can get cellphone coverage via a cellular network.

To enable 3G access, tap Settings, then General, and then tap Network. Tap to turn on the Enable 3G and Cellular Data settings. Do note that browsing the Internet using a 3G connection can eat up your data plan allotment quickly if your plan doesn't include unlimited data access.

To connect to a Wi-Fi network, you have to complete a few steps.

1. Tap the Settings icon on the Home screen and then tap Wi-Fi. Be sure Wi-Fi is set to On and choose a network to connect to. You should be able to access this automatically when you're in range of this network. When you're in range of a public hotspot, if access to several nearby networks is available, you may see a message asking you to tap a network name to select it. After you select one (or if only one network is available), you may see a message asking for your password. Ask the owner of the hotspot (for example a hotel desk clerk or business owner) for this password.

2. If you're required to enter a network password, do so.

3. Tap the Join button, and you're connected.

 Free public Wi-Fi networks typically don't require passwords. However, it's then possible for someone else to track your online activities over these *unsecured* networks. Avoid accessing financial accounts or sending sensitive e-mail when connected to a public hotspot.

Explore Safari

1. After you're connected to a network, tap the Safari icon on the Home screen. Safari opens, probably displaying the Apple iPhone home page the first time you go online (see **Figure 10-1**).

2. Put two fingers together on the screen and unpinch to enlarge the view. Double-tap the screen with a single finger to restore the default view size.

3. Put your finger on the screen and flick upward to scroll down on the page.

4. To return to the top of the web page, put your finger on the screen and drag downward or tap the Status bar at the top of the screen.

 Using the pinch method to enlarge or reduce the size of a web page on your screen allows you to view what's displayed at various sizes, giving you more flexibility than the double-tap method.

 When you enlarge the display, you gain more control using two fingers to drag from left to right or from top to bottom on the screen. On a reduced display, one finger works fine for making these gestures.

Address field   Status bar

Search field

Reload icon

Previous button   Page Navigation button

Next button | Bookmarks button

Menu button

Figure 10-1

Navigate among Web Pages

1. Tap in the Address field. The onscreen keyboard appears (see **Figure 10-2**).

2. To clear the field, press the Delete key on the keyboard. Enter a web address; for example, you can go to www.wiley.com.

3. Tap the Go key on the keyboard (refer to **Figure 10-2**). The website appears.

- If, for some reason, a page doesn't display, tap the Reload icon at the right end of the Address field.

- If Safari is loading a web page and you change your mind about viewing the page, you can tap the Stop icon (the X), which appears at the right end of the Address field during this process, to stop loading the page.

4. Tap the Previous arrow to go backward to the last page you displayed.

5. Tap the Next arrow to go forward to the page you just backed up from.

6. To follow a link to another web page, tap the link with your finger. To view the destination web address of the link before you tap it, just touch and hold the link; a menu appears that displays the address at the top, as shown in **Figure 10-3.**

Figure 10-2

The link's web address

Figure 10-3

 By default, AutoFill is turned on in iPhone, causing entries you make in fields such as the Address field to automatically display possible matching entries. You can turn off AutoFill by using iPhone Safari Settings.

Use Multi-Page Browsing

1. *Multi-page browsing* is a feature that allows you to have several websites open at once so you can move easily among those sites. With Safari open and a web page already displayed, tap the Page Navigation button (refer to **Figure 10-1**).

2. To add a new page (meaning you're opening a new website), tap the New Page button in the lower-left corner of the screen (see **Figure 10-4**). A new, blank page appears.

3. Tap in the address field, use the onscreen keyboard to enter the web address for the website you want to open, and then tap the Go key. The website opens on the page.

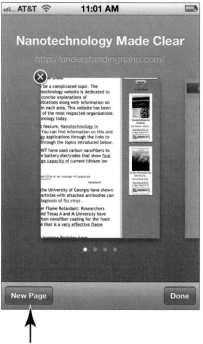

New Page button

Figure 10-4

 Repeat Steps 1–3 to open as many new web pages as you like.

4. You can now switch among open sites by tapping the Page Navigation button and scrolling among recent sites, finding the one you want and tapping on it.

 Using multi-page browsing you can make not only a site available but also a search results screen. If you recently searched for something, those search results will be on your Recent Searches list.

View Browsing History

 1. As you move around the web, your browser keeps a record of your browsing history. This record can be handy when you want to visit a site that you viewed previously but have forgotten its address. With Safari open, tap the Bookmark button.

2. On the menu shown in **Figure 10-5,** tap History.

3. In the History list that appears (see **Figure 10-6**), tap a date if available, and then tap a site to navigate to it.

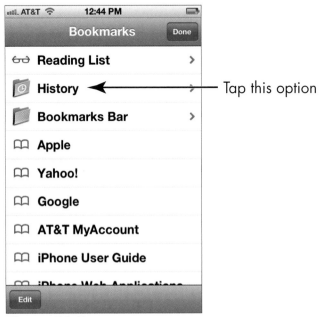

Tap this option

Figure 10-5

To clear the history, tap the Clear button (refer to **Figure 10-6**) and on the dialog that appears tap Clear History. This button is useful when you don't want your spouse or grandchildren to see where you've been browsing for birthday or holiday presents!

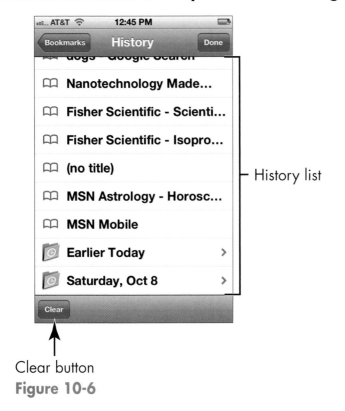

History list

Clear button

Figure 10-6

Search the Web

1. If you don't know the address of the site you want to visit (or you want to research a topic or find other information online), get acquainted with Safari's Search feature on iPhone. By default, Safari uses the Google search engine. With Safari open, tap in the Search field (refer to **Figure 10-1**). The onscreen keyboard appears.

2. Enter a search term. You can tap one of the suggested sites or complete your entry and tap the Search key (see **Figure 10-7**) on your keyboard.

Enter the search term here

Search results

Search key

Figure 10-7

3. In the search results that are displayed, tap a link to visit that site.

> To change your default search engine from Google to Yahoo! or Bing, in iPhone Settings, tap Safari and then tap Search Engine. Tap Yahoo! or Bing, and your default search engine changes.

> You can browse for specific items such as images, videos, or maps by tapping the corresponding link at the top of the Google results screen. Also, tap the More button in this list to see even more options to narrow your results, such as searching for books or YouTube videos on the subject.

Add and Use Bookmarks

 1. Bookmarks are a way to save favorite sites so you can easily visit them again. With a site you want to bookmark displayed, tap the Menu button.

2. On the menu that appears (see **Figure 10-8**), tap Add Bookmark.

— Tap this option

Figure 10-8

3. In the Add Bookmark dialog, shown in **Figure 10-9,** edit the name of the bookmark if you want. To do so, tap the name of the site and use the onscreen keyboard to edit its name.

Edit the name here

Figure 10-9

4. Tap the Save button.

 5. To go to the bookmark, tap the Bookmarks button.

6. On the Bookmarks menu that appears (see **Figure 10-10**), tap the bookmarked site you want to visit.

 If you want to sync your bookmarks on your iPhone browser to your computer, connect your iPhone to your computer and make sure that the Sync Safari Bookmarks setting on the Info tab of iTunes is activated. You can also go to Settings on iPhone and make sure iCloud is set to sync bookmarks.

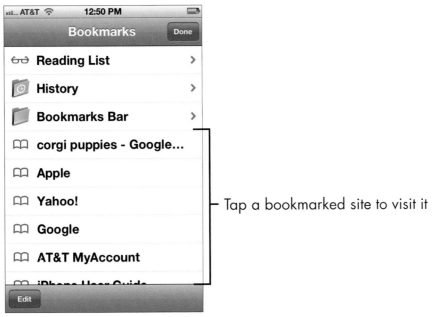

Figure 10-10

Tap a bookmarked site to visit it

When you tap the Bookmarks button, you can use the Bookmarks Bar option to create folders to organize your bookmarks. When you next add a bookmark, you can then choose, from the dialog that appears, any folder to which you want to add the new bookmark.

Use Safari Reading List

1. The Safari Reading List provides a way to save sites that contain content you want to read at a later time so you can easily visit them again. With a site you want to add to your Reading List displayed, tap the Menu button.

2. On the menu that appears (refer to **Figure 10-8**), tap Add to Reading List. The site is added to your list.

3. To view your Reading List, tap the Bookmarks button and tap Reading List.

4. On the Reading List that appears (see **Figure 10-11**), tap the content you want to revisit and resume reading.

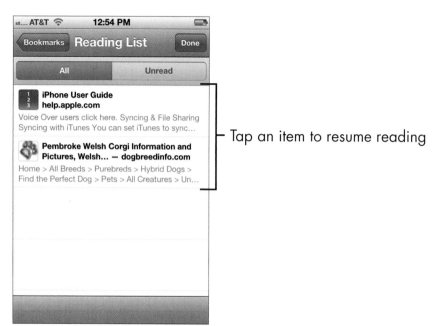

Tap an item to resume reading

Figure 10-11

If you want to see both Reading List material you've read and that material you haven't read, tap the All tab in the Reading List pane (refer to **Figure 10-11**). To see just the material you haven't read, use the Unread tab.

To save an image to your Reading List, tap and hold the image until a menu appears, and then tap Add to Reading list. To delete an item, with the Reading List displayed swipe left or right on an item, and a Delete

button appears. Tap it to delete the item from
Reading List.

Enjoy Reading More with Safari Reader

1. The Safari Reader feature gives you an e-reader type of
experience right within your browser, removing other sto-
ries and links as well as those distracting advertisements.
When you are on a site where you're reading content
such as an article, Safari displays a Reader button on the
right side of the address field (see **Figure 10-12**). Tap the
Reader button. The content appears in a reader format
(see **Figure 10-13**).

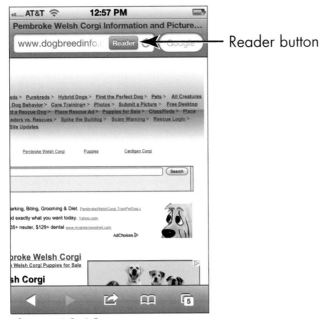

Reader button

Figure 10-12

Tap here to increase/decrease the text size

Figure 10-13

2. Scroll down the page. The entire content is contained in this one long page.

3. When you finish reading the material, just tap the Previous button to go back to its source.

 To enlarge the text in the Reader, tap the large A in the top-left corner (refer to **Figure 10-13**).

Add Web Clips to the Home Screen

 1. The Web Clips feature allows you to save a website as an icon on your Home screen so that you can go to the site at any time with one tap. With Safari open and displaying the site you want to add, tap the Menu button.

2. On the menu that appears (refer to **Figure 10-8**), tap Add to Home Screen.

3. In the Add to Home dialog that appears (see **Figure 10-14**), you can edit the name of the site to be more descriptive, if you like. To do so, tap the name of the site and use the onscreen keyboard to edit its name.

 ————— Edit the name here

Figure 10-14

4. Tap the Add button. The site is added to your Home screen.

 Remember that you can have as many as 11 Home screens on your iPhone to accommodate all the web clips you create and apps you download (though there is a limit to how many will fit). If you want to delete an item from a Home screen for any reason, press and hold the icon on the Home screen until all items on the screen start to jiggle and Delete icons appear on all items except the preinstalled apps. Tap the Delete icon on each item you want to delete, and it's gone. (To get rid of the jiggle, tap the Home button again.)

Save an Image to Your Photo Library

1. Display a web page that contains an image you want to copy.

2. Press and hold the image. The menu shown in **Figure 10-15** appears.

 ———— Tap this option

Figure 10-15

3. Tap the Save Image option (refer to **Figure 10-15**). The image is saved to your library.

 Be careful about copying images from the Internet and using them for business or promotional activities. Most images are copyrighted, and you may violate the copyright even if you simply use an image in (say) a brochure for your association or a flyer for your community group. Note that some search engines' advanced search settings offer the option of browsing only for images that aren't copyrighted.

Some websites are set up to prevent you from copying images on them, or display a pop-up stating that the contents on the site are copyrighted and should not be copied.

Send a Link

 1. If you find a great site that you want to share, you can do so easily by sending a link in an e-mail. With Safari open and the site you want to share displayed, tap the Menu button.

2. On the menu that appears (refer to **Figure 10-8**), tap Mail Link to This Page.

3. On the message form that appears (see **Figure 10-16**), enter a recipient's e-mail address, a subject, and your message.

Figure 10-16

4. Tap Send, and the e-mail goes on its way.

 The e-mail is sent from the default e-mail account you have set up on iPhone. For more about setting up an e-mail account, see Chapter 11.

 To tweet the link using your Twitter account, in Step 2 of this task choose Tweet, enter your tweet message in the form that appears, and then tap Send. For more about using Twitter with iPhone, see Chapter 9.

Make Private Browsing and Cookie Settings

Apple has provided some privacy settings for Safari that you should consider using. Private Browsing automatically removes items from the download list, stops Safari from letting AutoFill save information used to complete your entries in the search or address fields as you type, and doesn't save some browsing history information. These features can keep your online activities more private. The Accept Cookies setting allows you to stop the downloading of *cookies* (small files that document your browsing history so you can be recognized the next time you go to or move within a site) to your iPhone.

You can control both settings by choosing Safari in the Settings window. Tap to turn Private Browsing on or off (see **Figure 10-17**). Tap the arrow on Accept Cookies and choose to never save cookies, always save cookies, or only save cookies from visited sites.

 You can also tap the Clear History and the Clear Cookies and Data options (refer to **Figure 10-17**) to manually clear your browsing history, saved cookies, and other data.

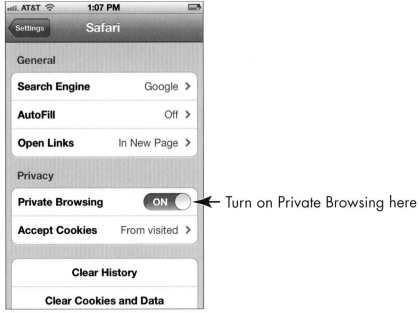

Turn on Private Browsing here

Figure 10-17

Print a Web Page

1. If you have a wireless printer that supports Apple's AirPrint technology (Hewlett-Packard is the only manufacturer that makes these at present), you can print web content using a wireless connection. With Safari open and the site you want to print displayed, tap the Menu button.

2. On the menu that appears (refer to **Figure 10-8**), tap Print.

3. In the Printer Options dialog that appears (see **Figure 10-18**), tap Select Printer. In the list of printers that appears, tap the name of your wireless printer.

Tap Select Printer

Figure 10-18

4. Tap either the plus or minus button in the Copy field to adjust the number of copies to print.

5. Tap Print to print the displayed page.

 The Mac applications Printopia and AirPrint Activator 2 make any shared or network printer on your home network visible to your iPhone. Printopia has more features, but is more expensive, while AirPrint Activator is free.

 If you don't have an AirPrint–compatible wireless printer or don't wish to use an app to help you print wirelessly, just e-mail a link to the web page to yourself, open the link on your computer, and print from there.

Working with E-mail in Mail

Staying in touch with others by using e-mail is a great way to use your iPhone. You can access an existing account using the handy Mail app supplied with your iPhone or sign in to your e-mail account using the Safari browser. Using Mail involves adding an existing e-mail account by way of iPhone Settings. Then you can use Mail to write, format, retrieve, and forward messages from that account.

Mail offers the capability to mark the messages you've read, delete messages, and organize your messages in a small set of folders, as well as a handy search feature. In this chapter, you read all about Mail and its various features.

Add an iCloud, Gmail, Yahoo!, AOL, or Windows Live Hotmail Account

1. You can add one or more e-mail accounts, including the e-mail account associated with your iCloud account, using iPhone Settings. If you have an iCloud, Gmail, Yahoo!, AOL, or Windows Live Hotmail account, iPhone pretty much automates the setup. To set up iPhone to retrieve messages from your e-mail account at one of these popular providers, first tap the Settings icon on the Home screen.

2. In the Settings dialog, tap Mail, Contacts, Calendars. The settings shown in **Figure 11-1** appear.

3. Tap Add Account. The options shown in **Figure 11-2** appear.

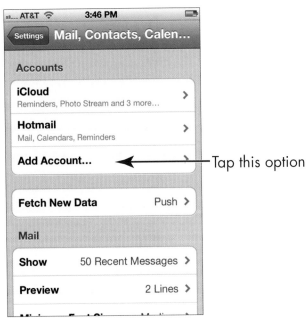

Tap this option

Figure 11-1

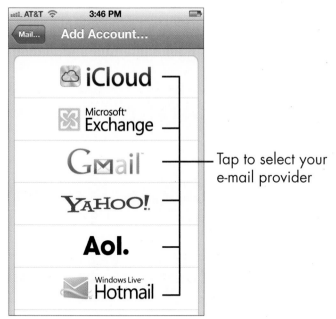

Tap to select your e-mail provider

Figure 11-2

4. Tap iCloud, Gmail, Yahoo!, AOL, or Windows Live Hotmail. Enter your account information in the form that appears (see **Figure 11-3**).

Enter account info here

Figure 11-3

5. After iPhone takes a moment to verify your account information, you can tap any On/Off button to have Mail, Contacts, Calendars, or Reminders from that account synced with iPhone.

6. When you're done, tap Save. The account is saved, and you can now open it using Mail.

Set Up a POP3 E-mail Account

1. You can also set up most popular e-mail accounts, such as those available through Earthlink or a cable provider's service, by obtaining the host name from the provider. To set up an existing account with a provider other than iCloud, Gmail, Yahoo!, AOL, or Windows Live Hotmail, you enter the account settings yourself. First, tap the Settings icon on the Home screen.

2. In Settings, tap Mail, Contacts, Calendars, and then tap the Add Account button (refer to **Figure 11-1**).

3. On the screen that appears (refer to **Figure 11-2**), scroll down and tap Other.

4. On the screen shown in **Figure 11-4**, tap Add Mail Account.

5. In the form that appears (refer to **Figure 11-3**), enter your name and an account address, password, and description, and then tap Next. iPhone takes a moment to verify your account and then returns you to the Mail, Contacts, Calendars page, with your new account displayed.

 If you have a less mainstream e-mail service, you may have to enter the mail server protocol (POP3 or IMAP — ask your provider for this information) and your password. iPhone will probably add the outgoing mail server (SMTP) information for you, but if it doesn't, you may have to enter it yourself. Your Internet service provider (ISP) can provide it to you.

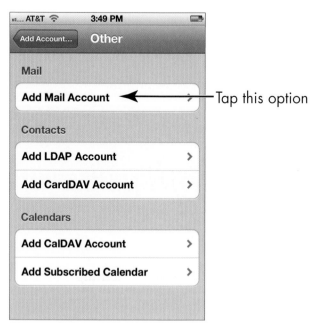

Tap this option

Figure 11-4

6. To make sure that the Account field is set to On for receiving e-mail, tap the account name. In the dialog that appears, tap the On/Off button for the Mail field and then tap the Mail button to return to Mail settings. You can now access the account through Mail.

 If you turn on Calendars in the Mail account settings, any information you've put into your calendar in that e-mail account will be brought over into the Calendar app on your iPhone and reflected in the Notifications Center (discussed in more detail in Chapter 21).

Open Mail and Read Messages

1. Tap the Mail app icon, located in the Dock on the Home screen (see **Figure 11-5**). A red circle on the icon indicates the number of unread e-mails in your Inbox.

2. In the Mail app, tap the Mailboxes button to display your inbox and folders (see **Figure 11-6**). Tap the inbox whose contents you want to display.

Tap this icon

Figure 11-5

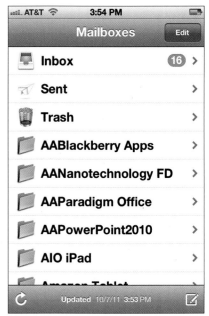

Figure 11-6

3. Tap a message to read it. It opens (see **Figure 11-7**).

4. If you need to scroll to see the entire message, just place your finger on the screen and flick upward to scroll down. You can also swipe right while reading a message to open the Inbox list of messages, and then swipe left to hide the list.

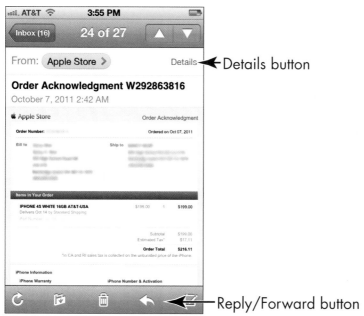

Figure 11-7

 You can tap the Hide button (top-right corner of the message) to hide the address details (the To field) so that more of the message appears on your screen. To reveal the field again, tap the Details button (which becomes the Hide button when details are displayed).

 E-mail messages you haven't read are marked with a blue circle in your Inbox. After you read a message, the blue circle disappears. You can mark a read message as unread, to help remind you to read it again later. With a message open and details about it displayed, tap the Mark link on the right side and then tap Mark as Unread. To flag a message, which places a little flag next to it in your inbox, helping you to spot items of more importance or to read again, tap Mark and then tap Flag.

 To escape your e-mail now and then (or to avoid having messages retrieved while you're at a public Wi-Fi hotspot), you can stop retrieval of e-mail by using the Fetch New Data control of Mail in iPhone Settings.

Reply to or Forward E-mail

1. With an e-mail message open (see the previous task), tap the Reply/Forward button (refer to **Figure 11-7**).

2. In the dialog that appears (see **Figure 11-8**), take one of the following actions:

- Tap Reply to respond to the sender of the message or Reply All to respond to the sender and any other recipients. The Reply Message form, shown in **Figure 11-9**, appears. Tap in the message body and enter a message.

Figure 11-8

Tap here to enter a reply

Figure 11-9

- Tap Forward to send the message to somebody else. If the original message had an attachment, you're offered the option of including or not including the attachment when forwarding.

The form shown in **Figure 11-10** appears. Enter a recipient in the To field, and then tap in the message body and enter a message.

Figure 11-10

3. Tap Send. The message goes on its way.

 If you want to move an address from the To field to the Cc or Bcc field, tap and hold the address and drag it to the other field.

Create and Send a New Message

 1. With Mail open, tap the New Message button. A blank message form appears (see **Figure 11-11**).

 2. Enter a recipient's address in the To field. If you have saved addresses in Contacts, tap the plus sign (+) in the Address field to choose an addressee from the Contacts list.

Figure 11-11

3. If you want to send a copy of the message to other people, enter their addresses in the Cc/Bcc field. If you want to send both carbon copies and *blind carbon copies* (copies you don't want other recipients to be aware of), note that when you tap the Cc/Bcc field, two fields are displayed; use the Bcc field to specify recipients for blind carbon copies.

4. Enter the subject of the message in the Subject field.

5. Tap in the message body and type your message.

6. Tap Send.

 Mail keeps a copy of all deleted messages for a time in the Trash folder. To view deleted messages, tap the Inbox button and then, in the dialog that appears, tap the Mailboxes button to show all mailboxes. If you have more than one mail account (therefore more than one mailbox), tap the account name in the Accounts list, and a list of folders opens. Tap the Trash folder, and all deleted messages are displayed.

Format E-mail

1. A new feature that comes with iOS 5 is the capability to apply formatting to e-mail text. You can use bold, underline, and italic formats, and indent text using the Quote Level feature.

2. Tap the text and choose Select or Select All to select a single word or all the words in the e-mail. Note that if you select a single word, handles appear that you can drag to add adjacent words to your selection.

3. Tap the arrow at the far end of the toolbar that appears.

4. To apply bold, italic, or underline formatting, tap the BIU button (see **Figure 11-12**).

5. In the toolbar that appears (see **Figure 11-13**), tap Bold, Italics, or Underline to apply formatting.

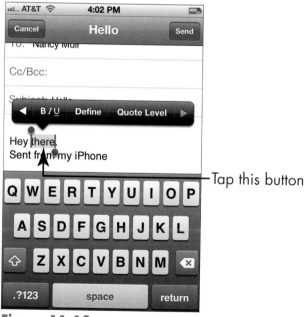

Tap this button

Figure 11-12

Make your selection here

Figure 11-13

6. To change the indent level, tap at the beginning of a line and then tap the arrow at the far end of the toolbar. In the toolbar that appears. tap Quote Level (refer to **Figure 11-12**).

7. Tap Increase to indent the text or Decrease to move indented text farther toward the left margin.

 To change the minimum size of text in e-mails, use the iPhone Settings. Tap Mail, Contacts, Calendars, and then tap the Minimum Font Size setting. Choose from Small, Medium, Large, Extra Large, and Giant in the list that appears.

Search E-mail

1. Say you want to find all messages from a certain person or containing a certain word in the Subject field. You can use Mail's handy Search feature to find these e-mails. With Mail open, tap the Inbox button.

2. In the Inbox, tap in the Search field. The onscreen keyboard appears.

3. Enter a search term or name as shown in **Figure 11-14**.

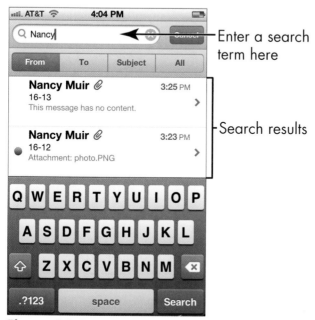

Figure 11-14

4. Tap the From, To, Subject tab to view messages that contain the search term in one of those fields, or tap the All tab to see messages in which any of these three fields contains the term. Matching e-mails are listed in the results (refer to **Figure 11-14**).

 You can also use the Spotlight Search feature covered in Chapter 2 to search for terms in the To, From, or Subject lines of mail messages.

 To start a new search or go back to the full Inbox, tap the Delete icon (the circled X) on the far-right end of the Search field to delete the term or just tap the Cancel button.

Delete E-mail

1. When you no longer want an e-mail cluttering your Inbox, you can delete it. With the Inbox displayed, tap the Edit button. Circular check boxes are displayed to the left of each message (see **Figure 11-15**).

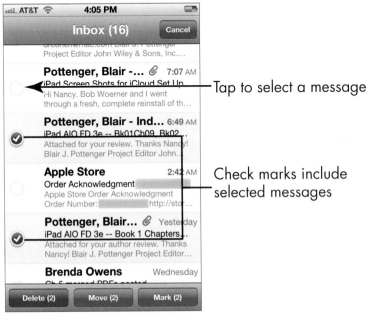

Tap to select a message

Check marks include selected messages

Figure 11-15

2. Tap the circle next to the message you want to delete. (You can tap multiple items if you have several e-mails to delete.) A message marked for deletion shows a check mark in the circular check button (refer to **Figure 11-15**).

3. Tap the Delete button at the bottom of the Inbox dialog. The message is moved to the Trash folder.

 You can also delete an open e-mail by tapping the trashcan icon on the toolbar that runs across the bottom of the screen, or swiping left or right on a message displayed in an inbox and tapping the Delete button that appears.

Organize E-mail

1. You can move messages into any of several predefined folders in Mail (these will vary depending on your e-mail provider and the folders you've created on your provider's server). After displaying the folder containing the message you want to move (for example, Trash or Inbox), tap the Edit button. Circular check boxes are displayed to the left of each message (refer to **Figure 11-15**).

2. Tap the circle next to the message you want to move.

3. Tap the Move button.

4. In the Mailboxes list that appears (see **Figure 11-16**), tap the folder where you want to store the message. The message is moved.

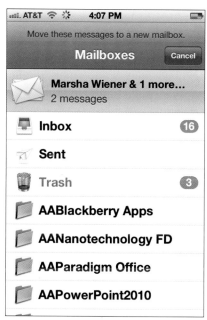

Figure 11-16

If you receive a junk e-mail, you might want to move it to the Spam or Junk folder if your e-mail account provides one. Then any future mail from the same sender is automatically placed in the Spam or Junk folder.

 If you have an e-mail open, you can move it to a folder by tapping the Folder icon on the toolbar along the top. The Mailboxes list displays; tap a folder to move the message.

Shopping the iTunes Store

The iTunes app that comes preinstalled in iPhone lets you easily shop for music, movies, TV shows, audiobooks, podcasts, and even online classes at Apple's iTunes Store.

In this chapter, you discover how to find content on the iTunes website. You can download the content directly to your iPhone or to another device and then sync it to your iPhone. In addition, I cover a few options for buying content from other online stores.

Note that I cover opening an iTunes account and downloading iTunes software to your computer in Chapter 3. If you need to, read Chapter 3 to see how to handle these two tasks before digging into this chapter.

Explore the iTunes Store

1. Visiting the iTunes Store from your iPhone is easy with the built-in iTunes app. Tap the iTunes icon on the Home screen.

2. If you're not already signed in to iTunes, the dialog shown in **Figure 12-1** appears, asking for your iTunes password. Enter your password and tap OK.

Tap here and enter your password

Figure 12-1

3. Tap the Music button (if it isn't already selected) in the row of buttons at the bottom of the screen.

4. Flick your finger up to scroll down the featured selections, as shown in **Figure 12-2.**

5. Tap the Top Tens tab at the top of the screen. This step displays best-selling songs and albums in the iTunes Store.

6. Tap any other listed item to see more detail about it, as shown in **Figure 12-3,** and hear a brief preview.

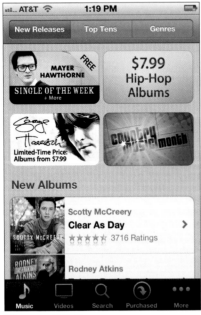

Figure 12-2

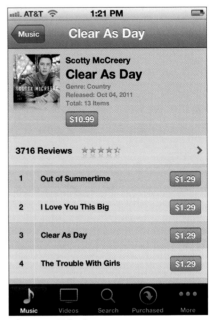

Figure 12-3

The navigation techniques in these steps work essentially the same in any of the content categories (the buttons at the bottom of the screen), which include Music, Videos, TV Shows, Podcasts, Audiobooks, and iTunes U (the last three are accessed through the More button at the bottom of the iTunes screen). Just tap one to explore it.

 If you want to use the Genius playlist feature, which recommends additional purchases based on the contents of your library in the iTunes app on your iPhone, tap the More button tab at the bottom of the screen and tap Genius. If you've made enough purchases in iTunes, song and album recommendations appear based on them as well as the content in your iTunes Match library, if you have one.

Find a Selection

You have several ways to look for a selection in the iTunes Store. You can use the Search feature, search by genre or category, or view artists' pages. Here's how these work:

➡ Tap in the Search field shown in **Figure 12-4** and enter a search term using the onscreen keyboard. Tap the Search button on the keyboard or, if a suggestion appeals to you, just tap that suggestion.

➡ Tap the Genres tab. (In some content types, such as Audiobooks, it's the Categories button; when you're viewing Video selections, you find three tabs across the screen for Movies, TV Shows, and Music Videos that take you to those categories.) A list of genres or categories like the one shown in **Figure 12-5** appears.

Enter a search term here

Tap here to view the list

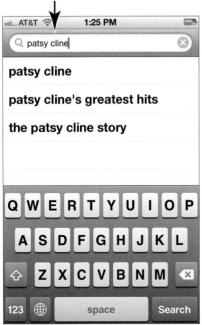

Figure 12-4

Figure 12-5

⟹ On a description page that appears when you tap a selection, you can find more offerings by the people involved with that particular work. For example, for a music selection, tap the More by This Artist link at the bottom of the page to see all of that artist's selections (see **Figure 12-6**).

Preview Music, a Video, or an Audiobook

1. Because you've already set up an iTunes account (if you haven't done so yet, refer to Chapter 3), when you choose to buy an item it's automatically charged to the credit card you have on record or against any allowance you have outstanding from an iTunes gift card. You might want to preview an item before you buy it. If you like it, buying and downloading are then easy and quick. Open iTunes and use any method outlined in earlier tasks to locate a selection you might want to buy.

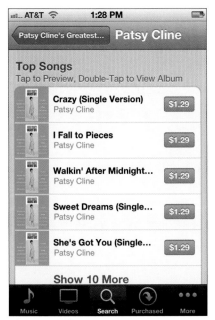

Figure 12-6

2. Tap the item to see detailed information about it, as shown in **Figure 12-7**.

3. If you're looking at a music selection, tap the track number or name of a selection (refer to **Figure 12-7**) to play a preview. For a movie or audiobook selection, tap the Preview button, shown in **Figure 12-8**.

 The iTunes Store offers several free selections, especially in the Podcast and iTunes U content categories (click the More button to find these categories). If you see one you want to try out, download it by tapping the button labeled Free and then tapping the Get Episode (or similarly named) button that appears.

Tap a track number or name to listen to a preview

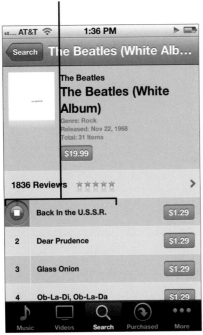

Figure 12-7

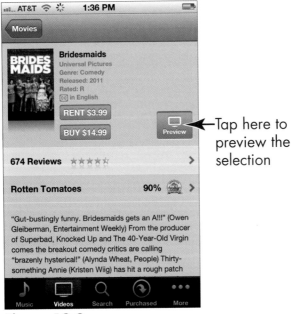

Tap here to preview the selection

Figure 12-8

Buy a Selection

1. When you find an item you want to buy, tap the button
that shows either the price (if it's a selection available for
purchase; see **Figure** 12-9) or the button with the word
Free on it (if it's a selection available for free).

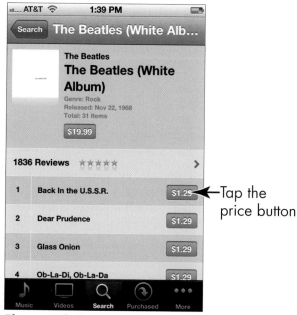

Tap the
price button

Figure 12-9

The button label changes to Buy X, where X is the type of
content, such as a song or album, you're buying.

2. Tap the Buy X button. The iTunes Password dialog
appears (refer to **Figure** 12-1).

3. Enter your password and tap OK. The item begins downloading, and the cost is automatically charged against
your account. When the download finishes, you can view
the content using the Music or Video app, depending on
the type of content.

If you want to buy music, you can open the description page for an album and buy individual songs rather than the entire album. Tap the price for a song, and then proceed to purchase it.

 Note the Redeem button on many iTunes screens. Tap this button to redeem any iTunes gift certificates you might get from your generous friends, or from yourself.

 If you don't want to allow purchases from within apps (for example Music or Videos) but rather want to allow purchases only through the iTunes store, you can go to Settings, General, tap Restrictions, and then tap Enable Restrictions, and enter a passcode. Once you've set a passcode, you can tap individual apps to turn on restrictions for them.

 You can allow content to be downloaded over your 3G cellular network if you're not near a Wi-Fi hotspot. Be aware, however, that this could incur hefty data charges with your provider if you run over your allotted data. However, if you aren't near a Wi-Fi hotspot, this might be your only option. Go to Settings, Store, and tap the On/Off button for the Cellular setting.

Rent Movies

1. In the case of movies, you can either rent or buy content. If you rent, which is less expensive, you have 30 days from the time you rent the item to begin to watch it. After you have begun to watch it, you have 24 hours from that time left to watch it as many times as you like. With iTunes open, tap the Videos button.

2. Locate the movie you want to rent and tap the arrow to the right of the movie, shown in **Figure 12-10**.

3. In the detailed description of the movie that appears, tap the Rent button, shown in **Figure 12-11**.

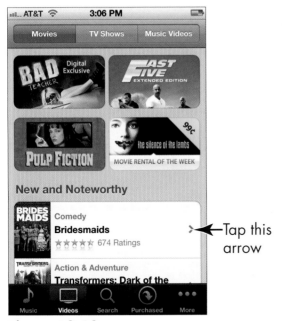

Figure 12-10

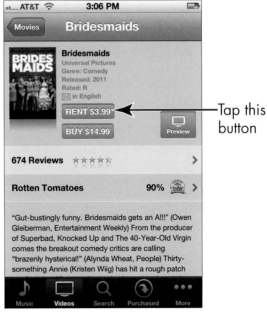

Figure 12-11

4. The gray Rent button changes to a green Rent Movie button; tap it to confirm the rental. The movie begins to download to your iPhone immediately, and your account is charged the rental fee.

5. To check the status of your download, tap the Downloads button. The progress of your download is displayed. After the download is complete, you can use the Videos app to watch it. (See Chapter 17 to read about how this app works.)

 Some movies are offered in high-definition versions. These HD movies look great on that crisp, colorful iPhone screen.

 You can also download content to your computer and sync it to your iPhone. Refer to Chapter 3 for more about this process.

Listen to Podcasts

1. *Podcasts* are audio or video broadcasts you can listen or watch to on your iPhone. Most of them are free, and iTunes features a wide variety of broadcast topics. With iTunes open, tap the More button in the row of buttons at the bottom of the screen and then tap Podcasts.

2. Tap a podcast selection; a detailed listing of podcasts, like the one shown in **Figure 12-12,** appears.

Figure 12-12

3. Tap the name or number of a podcast for additional information, or simply tap the Free button (refer to **Figure 12-12**), and then tap the Get Episode button to download the podcast. After it downloads, you can play it using the Music app.

 To save you work, you might consider using the subscription feature in iTunes. First, locate a podcast and click the Subscribe button. A URL for the podcast is displayed. Copy the URL, and then, with iTunes open, click the Advanced menu and choose Subscribe to Podcast. In the dialog that opens, enter the URL of a podcast you like. Now you'll get all episodes of that podcast delivered to iTunes as they become available, and they'll be available to all your devices through iCloud. (See more about setting up iCloud for automatically pushing purchases from one device to your other Apple devices in the final task of this chapter.)

Go to School at iTunes U

1. One very cool feature of iTunes is *iTunes U,* a compilation of free and often excellent online courses from universities and other providers. Tap More and then tap the iTunes U button in the list of other items to display selections.

2. Tap one of the three tabs shown in **Figure 12-13:** What's Hot, Top Tens, or Categories.

3. On the list that appears, tap an item to select the source for a course. That provider's page appears.

4. Tap either the Next or Previous button to scroll through offerings. When you find a topic of interest, tap a selection, and it opens, displaying a list of segments of the course, as shown in **Figure 12-14.**

5. Tap the Free button next to a course, and then tap the green Download/Get Episode button that appears. (If the course includes only one session, this button says Download, but if there's more than one session, it says Get Episode.) The course begins downloading.

After you're on a provider's page, to return to iTunes U, just tap the button labeled iTunes U (located in the top-left corner of the screen), and you return to the page with the three provider tabs (refer to **Figure 12-13**).

You can make selections of courses by tapping the Categories button and choosing a category of courses to browse, such as Fine Art or Business.

Tap one of these tabs

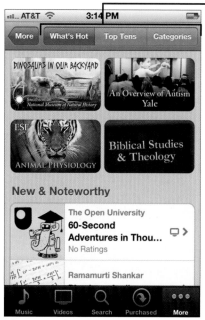

Figure 12-13

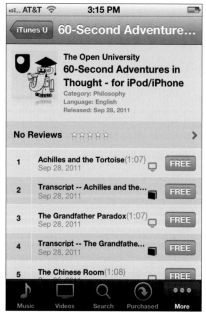

Figure 12-14

Shop Anywhere Else

One feature that's missing from the iPhone is support for Flash, a format of video playback that many online video-on-demand services use. However, many online stores that sell content such as movies and music are hurriedly adding iPhone-friendly videos to their collections, so you do have alternatives to iTunes for your choice of movies and TV shows. You can also shop for music from sources other than iTunes, such as Amazon.com.

You can open accounts at one of these stores by using your computer or your iPhone's Safari browser and then following the store's instructions for purchasing and downloading content.

These stores offer iPhone-compatible video content, and more are opening all the time:

⟹ **ABC:** http://abc.go.com

⟹ **CBS News:** www.cbsnews.com

⟹ **Clicker:** www.clicker.com

⟹ **Netflix:** www.netflix.com

⟹ **Ustream:** www.ustream.tv

 For non–iPhone-friendly formats, you can download the content on your computer and stream them to your iPhone using Air Video ($2.99) on the iPhone and Air Video Server (which is free) using your Mac or Windows computer. For more information go to www.inmethod.com/air-video/index.html.

Enable Auto Downloads of Purchases from Other Devices

1. With iCloud, you can make a purchase or download free content on any of your Apple devices, and iCloud

automatically shares those purchases with all your Apple
devices. To enable this auto download feature on iPhone,
start by tapping Settings on the Home screen.

 To use iCloud, first set up an iCloud account. See
Chapter 3 for detailed coverage of iCloud, including
setting up your account.

2. Tap Store.

3. In the options that appear, tap the On/Off button for any
category of purchases you want to auto download to your
iPhone from other Apple devices: Music, Apps, or Books
(see **Figure 12-15**).

Figure 12-15

 At this point, Apple doesn't offer an option of auto
downloading video content using these settings. You
can always download video directly to your iPhone
through the iTunes app or sync to your computer
using iTunes to get the content.

Expanding Your iPhone Horizons with Apps

*S*ome *apps* (short for *applications*) come pre-installed on your iPhone, such as Contacts and Videos. But you can choose from a world of other apps out there for your iPhone, some for free (such as iBooks) and some for a price (typically, from 99 cents to about $10 though some can top out at $90 or more).

Apps range from games to financial tools and productivity applications, such as the iPhone version of Pages, the Apple software for word processing and page layout.

In this chapter, I suggest some apps you might want to check out, and explain how to use the App Store feature of iPhone to find, purchase, and download apps.

Explore Senior-Recommended Apps

As I write this book, new iPhone apps are in development, so by the time you have this book in your hands, you'll find even more apps that can fit your needs. Still, to get you exploring what's available, I want to provide a quick list of apps that might whet your appetite.

Access the App Store by tapping the App Store icon on the Home screen. Then tap in the Search box and enter an app name. Suggested matches are listed. Tap one to see more information about it.

Here are some interesting apps to explore:

⮞ **Sudoku2 (Free):** If you like this mental math puzzle in print, try it out on your iPhone (see **Figure 13-1**). It has three lessons and several levels ranging from easiest to nightmare, making it a great way to make time fly by in the doctor's or dentist's waiting room.

Figure 13-1

⮞ **StockWatch Portfolio Tracking ($.99):** Keep track of your investments with this app on your iPhone. You can use the app to create a watch list and record your stock performance.

⮞ **Price Grabber iPhone Edition (Free):** Use this app to find low prices on just about everything. Read product reviews and compare list prices.

➡ **Flickr (Free):** If you use the Flickr photo-sharing service on your computer, why not bring the same features to your iPhone? This app is useful for sharing images with family and friends.

➡ **Paint Studio ($.99):** Get creative! You can use this powerful app to draw, add color, and even create special effects. If you don't need all these features, try Paint Studio Jr.

➡ **Virtuoso Piano Free 2 HD (Free):** If you love to make music, you'll love this app, which gives you a virtual piano keyboard to play and compose on the move.

➡ **Nike Training Club (Free):** Use this handy utility to help design personalized workouts, see step-by-step instructions to help you learn new exercises, and watch video demonstrations. The reward system in this app may just keep you going toward your workout goals.

 iBooks is the outstanding, free e-reader app that opens up a world of reading on your iPhone. See Chapter 14 for details about using iBooks.

Search the App Store

1. Tap the App Store icon on the Home screen. The site shown in **Figure 13-2** appears.

2. At this point, you have several options for finding apps:

- Tap the Search button at the bottom of the screen and then tap in the Search field, enter a search term, and tap the result you want to view.

- Tap the New, What's Hot, or Genius tab at the top of the screen to see that category of apps.

- Tap the Featured or Top 25 button at the bottom of the screen to see which free and paid apps other people are downloading most.

- Tap the Categories button to browse by type of app, as shown in **Figure** 13-3.

New, What's Hot, and Genius tabs

Featured button

Categories button

Top 25 button

Search button

Figure 13-2

Figure 13-3

Get Applications from the App Store

1. Buying apps requires that you have an iTunes account, which I cover in Chapter 3. After you have an account, you can use the saved payment information there to buy apps in a few simple steps or download free apps. I strongly recommend that you install the free iBooks app, so in this task, I walk you through the steps for getting it. With the App Store open, tap the Search button, enter iBooks in the Search field, and then tap the name of the app when it appears in the results list.

Note that when you first open the App Store, a message appears suggesting you download the iBooks app. If you already saw that message and downloaded it, feel free to download any free app in these steps, such as the Twitter app.

2. Tap the Free button for iBooks in the results that appear. (To get a *paid* app, you tap the same button, which is labeled with a price.)

3. The Free button changes to read *Install* (or, in the case of a paid app, the button changes to read *Buy*), shown in **Figure 13-4.** Tap the button; you may be asked to enter your iTunes password and tap the OK button to proceed.

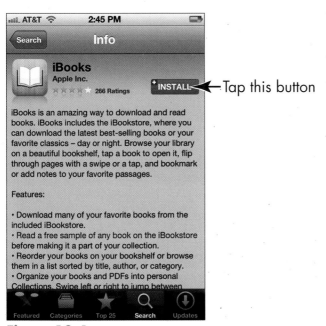

Figure 13-4

4. The app downloads; if you purchase an app that isn't free, at this point your credit card or gift card balance is charged for the purchase price.

Out of the box, only preinstalled apps are located on the first iPhone Home screen. Apps you download are placed on additional Home screens, and you have to scroll to view and use them; this procedure is covered later in this chapter. See the next task for help in finding your newly downloaded apps using multiple Home screens.

 If you've opened an iCloud account, you can set it up so that anything you purchase on your iPhone is automatically pushed to other Apple iOS devices and vice versa. See Chapter 3 for more about iCloud.

Organize Your Applications on Home Screens

1. iPhone can display up to 11 Home screens. By default, the first contains preinstalled apps, and the second Home screen contains a few more preinstalled apps in the Utilities folder; other screens are created to contain any apps you download or sync to your iPhone. At the bottom of any iPhone Home screen (just above the Dock), a magnifying-glass icon represents the Search screen to the left of the primary Home screen; dots that appear to the right of the magnifying-glass icon indicate the number of Home screens; and a solid dot specifies which Home screen you're on now, as shown in **Figure 13-5**. Tap the Home button to open the last displayed Home screen.

2. Flick your finger from right to left to move to the next Home screen. To move back, flick from left to right.

3. To reorganize apps on a Home screen, press and hold any app on that page. The app icons begin to jiggle (see **Figure 13-6**), and any apps you installed will sport a Delete button (a black circle with a white X on it).

4. Press, hold, and drag an app icon to another location on the screen to move it.

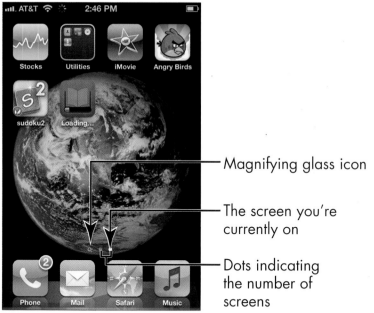

Magnifying glass icon

The screen you're currently on

Dots indicating the number of screens

Figure 13-5

A Delete button

Figure 13-6

5. Tap the Home button to stop all those icons from jiggling!

 To move an app from one page to another, while the apps are jiggling you can press, hold, and drag an app to the left or right to move it to the next Home screen. You can also manage what app resides on what Home screen from iTunes when you've connected iPhone to iTunes via a cable or wireless sync.

Organize Apps in Folders

iPhone lets you organize apps in folders. The process is simple:

1. Tap and hold an app until all apps do their jiggle dance.

2. Drag an app on top of another app. A bar appears across the screen, showing the two apps and a file with a placeholder name (see **Figure 13-7**).

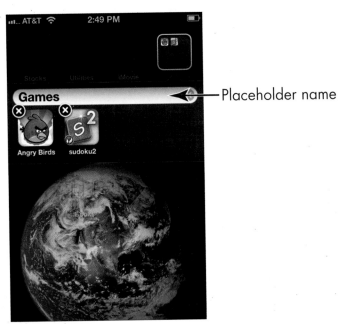

Figure 13-7

3. To change the name, tap in the field at the end of the placeholder name, and the keyboard appears.

4. Press the Delete key to delete the placeholder name and type one of your own.

5. Tap anywhere outside the bar to save the name.

6. Tap the Home key to stop the icons from dancing around and you'll see your folder appear on the Home screen where you began this process.

Delete Applications You No Longer Need

1. When you no longer need an app you have installed, it's time to get rid of it. (You can't delete apps that are preinstalled on the iPhone.) If you use iCloud to push content across all Apple iOS devices, note that deleting an app on your iPhone won't affect that app on other devices. Display the Home screen that contains the app you want to delete.

2. Press and hold the app until all apps begin to jiggle.

3. Tap the Delete button for the app you want to delete (see **Figure 13-8**).

4. A confirmation like the one shown in **Figure 13-9** appears. Tap Delete to proceed with the deletion.

5. A dialog asking you to rate an app before deleting it appears; you can tap the Rate button to rate it or tap No Thanks to opt out of the survey.

Don't worry about wiping out several apps at once by deleting a folder. When you delete a folder, the apps that were contained within the folder are placed back on the Home screen.

Delete buttons

Figure 13-8

Tap this button

Figure 13-9

Update Apps

1. App developers update their apps all the time, so you might want to check for those updates. Assuming Notifications Settings are set up to alert you to updates, the App Store icon on the Home screen will display the number of available updates. If you have turned this setting on, tap the App Store icon on the Home Screen. At the bottom right of the App Store screen is an Updates button, which will indicate the number of updates available, if any, in a red circle.

2. Tap the Updates button to access the Updates screen (see **Figure 13-10**) and then tap any item you want to update. To update all, tap the Update All button.

←Updates button

Figure 13-10

3. On the app screen that appears, tap Update. You may be asked to confirm that you want to update, or to enter your Apple ID and then tap OK to proceed. You may also be asked to confirm that you are over a certain age or agree to terms and conditions; if so, scroll down the terms dialog and, at the bottom, tap Agree.

 In iOS 5, Apple introduced the capability to download multiple apps at once. If you choose more than one app to update instead of downloading them sequentially as in previous versions of the iOS, all items will download simultaneously.

 If you have an iCloud account and update an app on your iPhone, it also updates on any other Apple iOS devices automatically and vice versa.

Part IV

Having Fun and Consuming Media

The 5th Wave   By Rich Tennant

"You should see the detail in the Satellite view of this area. It's like we're standing right there."

Using Your iPhone as an E-reader

A traditional *e-reader* is a device that's used to read the electronic version of books, magazines, and newspapers. Apple has touted iPhone as a great e-reader, although it isn't a traditional e-reader device like the Barnes & Noble Nook because it gets its functionality from an e-reader app.

Apple's free, downloadable app that turns your iPhone into an e-reader is *iBooks*, which also enables you to buy and download books from Apple's iBookstore. You can also use one of several other free e-reader apps — such as Kindle, Stanza, or Nook — to download books to your iPhone from a variety of online bookstores and other sources such as Google so you can read to your heart's content.

 A new app that comes with iOS 5 is Newsstand, covered later in this chapter. Newsstand has a similar look and feel to iBooks, but its focus is on subscribing to and reading magazines, newspapers, and other periodicals.

In this chapter, you discover the options available for reading material and how to buy books

Get ready to . . .

and subscribe to publications. You also learn all about Newsstand and iBooks: how to navigate a book or periodical and adjust the brightness and type, as well as how to search books and organize your iBooks and Newsstand libraries.

Discover E-reading

An *e-reader* is any electronic device that enables you to download and read books, magazines, or newspapers. These devices are typically portable and dedicated only to reading the electronic version of published materials. Many e-readers use E Ink technology to create a paperlike reading experience.

The iPhone is a bit different: It isn't only for reading books, and you have to download an app to enable it as an e-reader (though the apps are free). Also, the iPhone doesn't offer the paperlike reading experience — you read from a phone screen (though you can adjust the brightness and background color of the screen).

When you buy a book online (or get one of many free publications), it downloads to your iPhone in a few seconds using a Wi-Fi or 3G connection. The iPhone offers several navigation tools to move around a book, which you explore in this chapter.

Find Books with iBooks

1. In Chapter 13, I walk you through the process of downloading the iBooks application in the "Get Applications from the App Store" task, so you should do that first, if you haven't already. To shop using iBooks, tap the iBooks application icon to open it. (It's probably on your second Home screen, so you may have to swipe your finger to the left on the Home screen to locate it.)

2. In the iBooks library that opens (see **Figure 14-1**), you see a bookshelf; yours probably has only one free book already downloaded to it. (If you don't see the bookshelf, tap the Library button to go there.) Tap the Store button, and the shelf pivots 180 degrees.

Tap the Store button

Figure 14-1

3. In the iBookstore, shown in **Figure 14-2,** featured titles are shown by default. You can do any of the following to find a book:

- Tap the Search button at the bottom of the screen and type a search word or phrase in the search field that appears, using the onscreen keyboard.

- Tap the Categories button at the top of the screen to see a list of types of books, as shown in **Figure 14-3.** Tap a category to view those selections.

- Press your finger on the screen and flick up to scroll to more suggested titles on a page.

- Tap the appropriate button at the bottom of the screen to view particular categories: featured titles, Charts to see books listed on top charts, browse-worthy lists of authors or categories or only paid or free items, or titles you've already purchased.

If you go to an item by tapping a button at the bottom of the screen and you want to return to the original screen, just tap Featured again.

• Tap a suggested selection or featured book to read more information about it.

Categories button

Library button

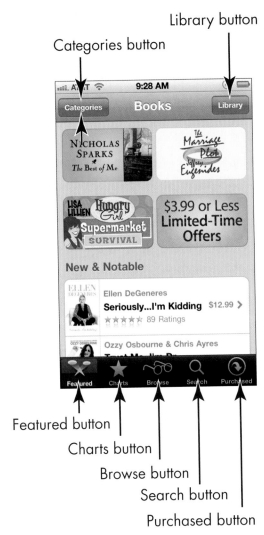

Featured button

Charts button

Browse button

Search button

Purchased button

Figure 14-2

Figure 14-3

Download free samples before you buy. You get to read several pages of the book to see whether it appeals to you, and it doesn't cost you a dime! Look for the Get Sample button when you view details about a book.

Explore Other E-book Sources

The iPhone is capable of using iBooks and other e-reader apps to read book content from other bookstores, so you can get books from sources other than iBookstore. To do so, first download another e-reader application such as Kindle from Amazon or the Barnes & Noble Nook reader from the iPhone App Store. (See Chapter 13 for how to download apps.) Then use their features to search for, purchase, and download content.

The Kindle e-reader application is shown in **Figure 14-4.** Any content you've already bought from the Amazon.com Kindle Store is archived online and can be placed on your Kindle Home page on the iPhone for you to read anytime you like. To delete a book from this reader, press the title with your finger, and the Delete button appears.

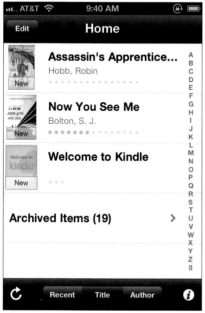

Figure 14-4

You can also get content from a variety of other sources: Project Gutenberg, Google, some publishers like Baen, and so on. Get the content using your computer if you like and then just add the items to Books in iTunes and sync them to your iPhone. You can also make settings to iCloud so that books are pushed across your Apple devices or place them in an online storage service such as Dropbox and access them from there.

Buy Books

1. If you've set up an account with iTunes, you can buy books at the iBookstore easily. (See Chapter 3 for more about setting up an account.) When you find a book in the iBookstore that you want to buy, tap its Price button. The button changes to the Buy Book button, as shown in **Figure 14-5.** (If the book is free, these buttons are labeled Free and Get Book respectively.)

2. Tap the Buy Book or Get Book button. If you haven't already signed in, the iTunes Password dialog, shown in **Figure 14-6,** appears.

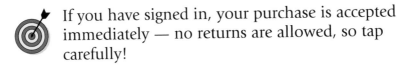 If you have signed in, your purchase is accepted immediately — no returns are allowed, so tap carefully!

3. Enter your password and tap OK.

4. The book appears on your bookshelf, and the cost is charged to whichever credit card you specified when you opened your iTunes account.

 You can also sync books you've downloaded to your computer to your iPhone by using the Dock Connector to USB Cable and your iTunes account or by using the wireless iTunes Wi-Fi Sync setting on the General Settings menu. Using this method, you can find lots of free books from various sources online and drag them into your iTunes Book library; then simply sync them to your iPhone. See Chapter 3 for more about syncing.

Buy Book button

Figure 14-5

Figure 14-6

Navigate a Book

1. Tap iBooks and, if your Library (the bookshelf) isn't already displayed, tap the Library button.

2. Tap a book to open it. The book opens to its title page or the last spot you read, as shown in **Figure 14-7**.

3. Take any of these actions to navigate the book:

- **To go to the book's Table of Contents:** Tap the Table of Contents button at the top of the page (refer to **Figure 14-7**) and then tap the name of a chapter to go to it (see **Figure 14-8**).

Table of Contents button

Slider to move to another page

Figure 14-7

Tap any chapter to go to it

Figure 14-8

- **To turn to the next page:** Place your finger anywhere along the right edge of the page and flick to the left.

- **To turn to the preceding page:** Place your finger anywhere on the left edge of a page and flick to the right.

- **To move to another page in the book:** Tap and drag the slider at the bottom of the page (refer to **Figure 14-7**) to the right or left.

 To return to the Library to view another book at any time, tap the Library button. If the button isn't visible, tap anywhere on the page, and the button and other tools appear.

Adjust Brightness in iBooks

1. iBooks offers an adjustable brightness setting that you can use to make your book pages comfortable to read. With a book open, tap the Brightness button, shown in **Figure** 14-9.

Tap the Brightness button...

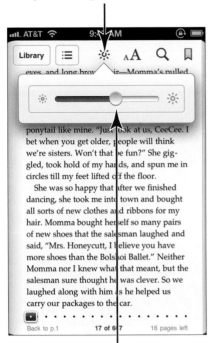

then adjust the screen brightness

Figure 14-9

2. On the Brightness setting that appears (refer to **Figure 14-9**), tap and drag the slider to the right to make the screen brighter, or to the left to dim it.

3. Tap anywhere on the page to close the Brightness dialog.

 Experiment with the brightness level that works for you, or try out the Sepia setting in the Fonts dialog. Bright white screens are commonly thought to be hard on the eyes, so setting the brightness halfway to its default setting (or less) is probably a good idea.

Change the Font Size and Type

1. If the type on your screen is a bit small for your taste, you can change to a larger font size or choose a different font for readability. With a book open, tap the Font button (it sports a small letter *a* and a large capital *A*, as shown in **Figure 14-10**).

2. In the Font dialog that appears (refer to **Figure 14-10**), tap the button with a small A on the left to use smaller text, or the button with the large A on the right to use larger text.

3. Tap the Fonts button. The list of fonts shown in **Figure 14-11** appears.

Tap the Font button

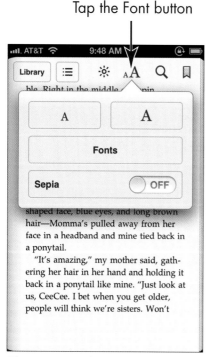

Figure 14-10   **Figure 14-11**

4. Tap a font name to select it. The font changes on the book page.

5. If you want a sepia tint on the pages, which can be easier on the eye, tap the On/Off button to turn on the setting.

6. Tap outside the Fonts dialog to return to your book.

 Some fonts appear a bit larger on your screen than others because of their design. If you want the largest fonts, use Cochin or Verdana.

Search in Your Book

1. You may want to find a certain sentence or reference in your book. To do so, with the book displayed, tap the Search button shown in **Figure 14-12.** The onscreen keyboard appears.

2. Enter a search term and then tap the Search key on the keyboard. iBooks searches for any matching entries.

3. Use your finger to scroll down the entries (see **Figure 14-13**).

Search button

Scroll through the entries

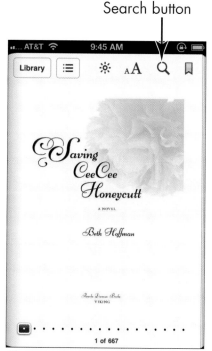

Figure 14-12

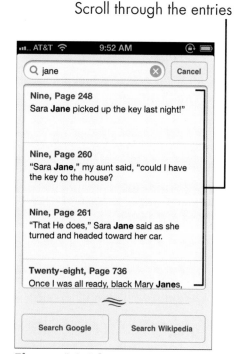

Figure 14-13

4. You can use either the Search Google or Search Wikipedia button at the bottom of the Search dialog if you want to search for information about the search term online.

 You can also search for other instances of a particular word while in the book pages by pressing your finger on the word and tapping Search on the toolbar that appears.

Use Bookmarks and Highlights

1. Bookmarks and highlights in your e-books are like favorite sites you save in your web browser: They enable you to revisit a favorite passage or refresh your memory about a character or plot point. To bookmark a page, with that page displayed just tap the Bookmark button in the top-right corner (see **Figure 14-14**).

2. To highlight a word or phrase, press a word until the toolbar shown in **Figure 14-15** appears.

3. Tap the Highlight button. A colored highlight is placed on the word.

4. To change the color of the highlight, add a note, or remove the highlight, tap the highlighted word. The toolbar shown in **Figure 14-16** appears.

5. Tap one of these three buttons:

- *Colors:* Displays a menu of colors you can tap to change the highlight color.

- *Note:* Lets you add a note to the item.

- *Remove Highlight:* Removes the highlight.

Bookmark button   Tap this button

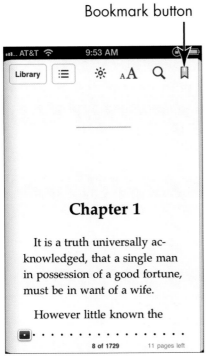

Figure 14-14

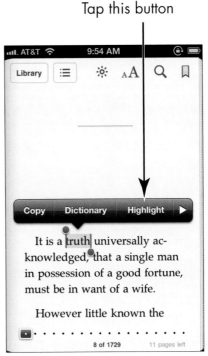

Figure 14-15

6. Tap outside the highlighted text to close the toolbar.

7. To go to a list of bookmarks and highlights, tap the Table of Contents button (refer to **Figure 14-7**) on a book page.

8. In the Table of Contents, tap the Bookmarks tab. As shown in **Figure 14-17,** all bookmarks are displayed.

9. Tap a bookmark or highlight in this list to go there.

 iPhone automatically bookmarks where you left off reading in a book so you don't have to mark your place manually. If you use any other device registered to your account you can also pick up where you left off reading.

Bookmarks and Highlights are
displayed here

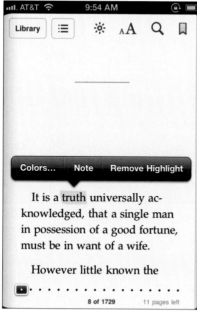

Figure 14-16

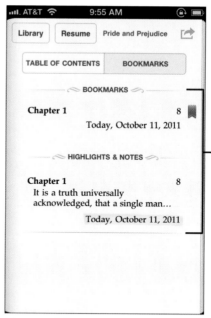

Figure 14-17

 You can also highlight illustrations in a book.
Display the page and press the image until the
Highlight button appears above it. Tap the button,
and the illustration is highlighted in yellow. As with
highlighted text, you can tap a bookmarked illustration to change the highlight color, add a note, or
remove its highlight.

Check Words in the Dictionary

1. As you read a book, you may come across unfamiliar words.
Don't skip over them — take the opportunity to learn a
word! With a book open, press your finger on a word and
hold it until the toolbar shown in **Figure 14-18** appears.

2. Tap the Dictionary button. A definition dialog appears,
as shown in **Figure 14-19**.

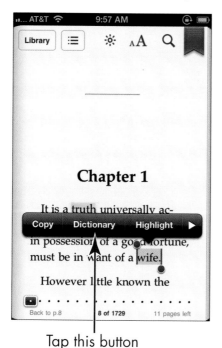

Tap this button

Figure 14-18

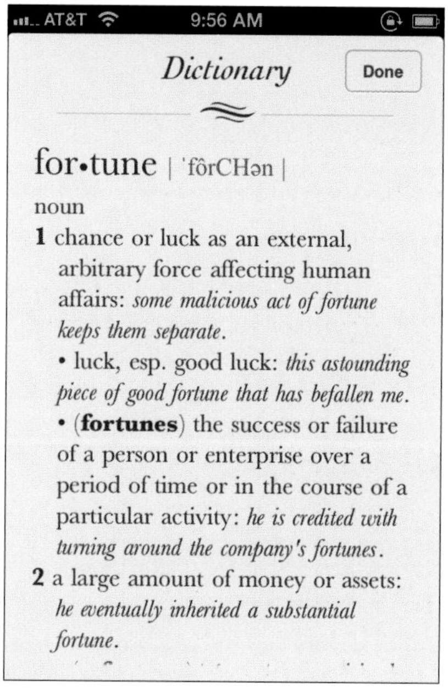

Figure 14-19

3. Tap the definition and scroll down to view more.

4. When you finish reviewing the definition, tap Done, and the definition disappears.

Organize Books in Collections

1. iBooks lets you create collections of books to help you organize them by your own logic, such as Tear Jerkers, Work-related, and Great Recipes. You can place a book in only one collection, however. To create a collection from the Library bookshelf, tap Collections (it's labeled with the currently displayed collection, probably Books by default).

2. In the dialog that appears, tap New. On the blank line that appears, type a name (see **Figure 14-20**).

3. Tap Done, which closes the dialog and returns you to the Library. To add a book to a collection from the Library, tap Edit.

4. Tap a book and then tap the Move button that appears in the top-left corner of the screen. In the dialog that appears, tap the collection to which you'd like to move the book, and the book now appears on the bookshelf in that collection (see **Figure 14-21**).

Enter a name here...

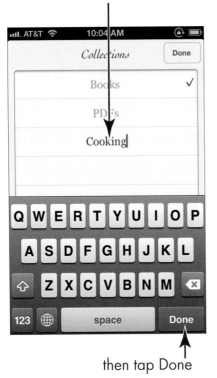

then tap Done

Figure 14-20

Figure 14-21

5. To delete a book from a collection with the collection displayed, tap Edit, tap the selection circle for the book, and then tap Delete.

 To delete a collection with the Collections dialog displayed, select the collection you want to delete and then tap Edit. Tap the minus sign to the left of the collection and then tap Delete to get rid of it. A message appears, asking you to tap Remove to remove the contents of the collection from your iPhone or Don't Remove. Note that if you choose Don't Remove, all titles within a deleted collection are returned to their original collections in your library.

Download Magazine Apps to Newsstand

1. Newsstand is a new app that focuses on subscribing to and reading magazines, newspapers, and other periodicals rather than books. The app has a similar look and feel to iBooks. When you download a free publication, you're actually downloading an app to Newsstand. You can then tap that app to buy individual issues, as covered in the next section. Tap the Newsstand icon on the Home screen to open Newsstand (see **Figure 14-22**).

2. Tap the Store button (refer to **Figure 14-22**). The store opens, displaying Featured periodicals; there are also Featured, Categories, and Top 25 icons along the bottom of the screen but these take you to other kinds of content (see **Figure 14-23**).

Tap this button

Figure 14-22

Figure 14-23

3. Tap any of the items displayed, scroll down the screen to view other choices, or tap the Search button on the bottom of the screen and enter a search term to locate a publication you're interested in.

 If you tap other icons at the bottom of the screen, such as Categories or Top 25, you're taken to other types of content than periodicals. Also, if you tap the Featured icon again after tapping one of these icons, you're taken to other kinds of apps than periodicals. If you're focused on finding only periodicals, your best bet is to stay on the Store screen that displays when you tap the Store button in Newsstand.

4. When you find an item, tap it to view a detailed description (see **Figure 14-24**).

Tap this button to install the app

Figure 14-24

5. Tap the Free button and then tap Install App. The app downloads to Newsstand.

Preview and Buy Issues of Periodicals through Newsstand

1. Tap a periodical app that you've added to Newsstand. The message shown in **Figure 14-25** appears, asking if you'd like to be informed of new issues. Tap OK if you would.

2. Tap the Preview Issue button to take a look at a description of its content, or tap the Buy button.

Tap this button

Figure 14-25

3. In the purchase confirmation dialog that appears, tap Buy. The issue is charged to your iTunes account.

 Note that different stores offer different options for subscribing, buying issues, and organizing issues. Think of Newsstand as a central collection point for apps that allow you to preview and buy content in each publication's store.

Read Periodicals in Your Newsstand Apps

1. If you buy a periodical or you've downloaded a free sub-scription preview, such as *The New York Times* Update shown in **Figure 14-26,** you can tap the publication in Newsstand to view it.

2. In the publication that appears, use your finger to swipe left, right, up, and down to view more of the pages.

3. In many publications, you can tap the Sections button (refer to **Figure 14-26)** to view the publication's Table of Contents, as shown in **Figure 14-27.**

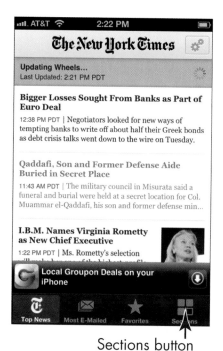

Sections button

Figure 14-26

Figure 14-27

4. Tap on a section or article name in the Table of Contents to view and read it.

Playing with Music on iPhone

*Y*ou've probably heard of the iPod — a small, portable, music-playing device from Apple that's seemingly glued into the ears of many kids and teens. iPhone includes an iPod-like app called Music that allows you to take advantage of its amazing little sound system to play your own style of music or podcasts and audiobooks. In this chapter, you get acquainted with the Music app and its features that allow you to sort and find music and control playback. You also get an overview of the Ping feature of iTunes, which shares your musical preferences with others, and AirPlay for accessing and playing your music over a home network.

Chapter 15

Get ready to . . .

View the Library Contents

1. Tap the Music app icon, located in the Dock on the Home screen. The Music library appears (the Albums view is shown **Figure 15-1**).

2. Tap the Playlists, Songs, Artists, or Albums buttons at the bottom of the library to view your music according to these criteria (refer to **Figure 15-1**).

3. Tap the More button (see **Figure 15-2**) to view music by genre or composer, or to view any audiobooks you've acquired.

Tap a criteria to use

Figure 15-1

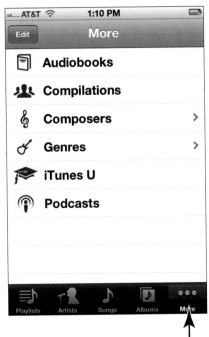

Tap this button

Figure 15-2

 iTunes has several free items you can download and use to play around with the features in Music, including music and podcasts. You can also sync content such as iTunes Smart Playlists stored on your computer or other Apple devices to your iPhone, and play it using the Music app. (See Chapter 3 for more about syncing and Chapter 12 for more about getting content from iTunes.)

Apple offers a service called iTunes Match (visit `www.apple.com/itunes/whats-new` for more information). You pay $24.99 per year for the capability to match the music you've bought from other providers (and stored on your computer) to what's in the iTunes library. If there's a match (and there usually is), that content is added to your iTunes library. Then, using iCloud, you can sync the content among all your Apple devices.

Create Playlists

1. You can create your own playlists to put tracks from various sources into collections of your choosing. Tap the Playlists button at the bottom of the iPhone screen.

2. Tap Add Playlist. In the dialog that appears, enter a name for the playlist and tap Save.

3. In the list of selections that appears (see **Figure 15-3**), tap the plus sign next to each item you want to include.

4. Tap the Done button, and then tap the Playlists button to return to the Playlists screen.

5. Your playlist appears in the list, and you can now play it by tapping the list name and then tapping a track to play it.

To search for a song in your music libraries, use the Spotlight Search feature. From your first Home screen, you can swipe to the right (or tap the first magnifying-glass icon to the left on any home screen) and enter the name of the song. It should appear in a list of search results.

Tap the plus sign to include
a song in the playlist

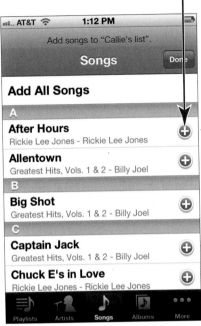

Figure 15-3

Search for Audio

1. You can search for an item in your Music library by using
the Search feature. With Music open, tap the Songs button and then flick downward on the screen to reveal the
Search field.

2. Tap in the Search field that appears at the top of the
screen (see **Figure** 15-4) to open the onscreen keyboard.

3. Enter a search term in the Search field. Results are displayed, narrowing as you type, as shown in **Figure** 15-5.

4. Tap an item to play it.

 You can enter an artist's name, an author's or a composer's name, or a word from the item's title in the
Search field to find what you're looking for.

Tap here to search for audio

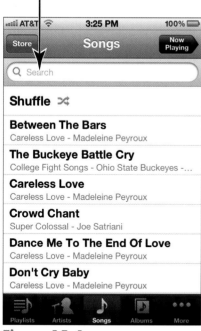

Figure 15-4

Results display here

Figure 15-5

Play Music and Other Audio

1. Locate the song or audiobook you want to play using the methods described in previous tasks in this chapter.

2. Tap the item you want to play. Note that if you're displaying the Songs tab, you don't have to tap an album to open a song; you need only tap a song to play it. If you're using any other tab, you have to tap items such as albums (or multiple songs from one artist) to find the song you want to hear.

3. Tap the item you want to play from the list that appears (see **Figure 15-6**); it begins to play.

4. Use the Previous and Next buttons at the bottom of the screen shown in **Figure 15-7** to navigate the audio file that's playing. The Previous button takes you back to the beginning of the item that's playing; the Next button takes you to the next item. Use the Volume slider on the bottom of the screen (or the Volume buttons on the side of your iPhone) to increase or decrease the volume.

5. Tap the Pause button to pause playback. Tap the button again to resume playing.

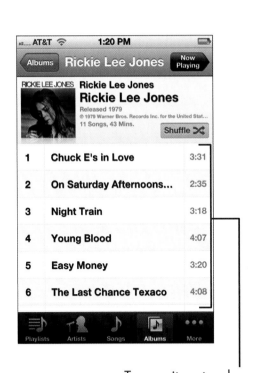

Tap an item to play

Figure 15-6

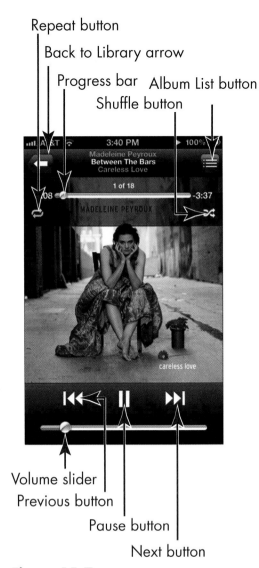

Repeat button

Back to Library arrow

Progress bar Album List button

Shuffle button

Volume slider
Previous button
Pause button
Next button

Figure 15-7

6. Tap and drag the slider that indicates the current play-back location on the Progress bar to the left or right to "scrub" to another location in the song. If the slider isn't visible, tap the image of the album cover, and the Progress bar appears.

7. If you don't like what's playing, here's how to make another selection: Tap the Back to Library arrow in the top-left corner to return to Library view or tap the Album List button in the bottom-right corner to show other selections in the album that's playing.

Shuffle Music

1. If you want to play a random selection of the music you've purchased or synced to your iPhone, you can use the Shuffle feature. With Music open, tap the Songs button at the bottom of the screen.

2. Tap the Shuffle button (see **Figure 15-8**). Your content plays in random order.

 If you're playing music and have set the Volume slider as high as it goes and you're still having trouble hearing, consider getting a headset. It cuts out extraneous noises and may improve the sound quality of what you're listening to, as well as adding stereo to iPhone's mono speaker. Preferably, you should use a 3.5mm stereo headphone; insert it in the headphone jack at the top of your iPhone.

Tap this button

Figure 15-8

Understand Ping

Ping is a social network for music lovers. If you think of Facebook or other social networking sites and imagine that they're focused around musical tastes, you have a good image of what Ping is all about.

After you join Ping, you can share the music you like with friends, take a look at the music your friends are purchasing, and follow certain artists (see **Figure 15-9**). You can also access short previews of the sounds your friends like.

If this sounds like something your grandchildren would do, you might be right. But if you have a few musical friends with whom you want to connect, Ping can be great fun.

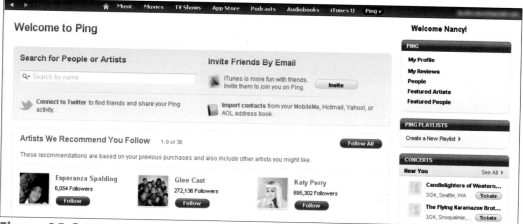

Figure 15-9

To use Ping, you need to activate it using iTunes on a PC or Mac by clicking the Ping link in the Source List and clicking the Turn On Ping button. You're then asked to fill in profile information, such as your name, gender, town, and musical preferences (see **Figure 15-10**). After you save this information, you have a Ping home page and can start inviting friends to share musical inspirations with you.

Figure 15-10

One cool feature of Ping is that, on your Ping page, a list of concerts near you is displayed. Even in my small town, Ping alerted me to some interesting events to check out.

Use AirPlay

The AirPlay streaming technology is built into the iPhone, iPod touch, and iPad. *Streaming* technology allows you to send media files from one device to be played on another. You can send (say) a movie you've purchased on your iPhone or a slideshow of your photos to be played on your TV — and control the TV playback from your iPhone. You can also send music to be played over speakers.

You can take advantage of AirPlay in a few ways: Purchase Apple TV and stream video, photos, and music to the TV or purchase AirPort Express and attach it to your speakers to play music. Finally, if you buy AirPort Express you can stream audio directly to your wireless speakers. Because this combination of equipment varies, my advice — if you're interested in using AirPlay — is to visit your nearest Apple Store and find out which hardware combination will work best for you.

 If you get a bit antsy watching a long movie, one of the beauties of AirPlay is that you can still use your iPhone to check e-mail, browse photos or the Internet, or check your calendar while the media file is playing.

Playing with Photos

With its gorgeous screen, the iPhone is a natural for viewing photos. It supports most common photo formats, such as JPEG, TIFF, and PNG. You can shoot your photos by using the built-in cameras in iPhone or sync photos from your computer. You can also save images you find online to your iPhone or receive them by e-mail.

When you have photos to play with, the Photos app lets you organize photos from the Camera Roll, view photos in albums, one by one, or in a slideshow. You can also e-mail, message, or tweet a photo to a friend or print it. You can read about all these features in this chapter.

Get ready to . . .

Take Pictures with the iPhone Cameras

1. The cameras in the iPhone 4 and 4S are just begging to be used, so let's get started! Tap the Camera app icon on the Home screen to open the app.

2. If the Camera/Video slider setting at the bottom-right corner of the screen (see **Figure 16-1**) is shifted to the right, slide it to the left to choose the still camera rather than video.

 iPhone's front- and rear-facing cameras allow you to capture photos and video (see Chapter 17 for more about the video features) and share them with family and friends. The 4S iPhone sports 8mp and 1080p cameras; the iPhone 4 provides a 5mp and 720p camera.

3. Tap the Options button, tap the On/Off button for Grid, and then tap Done. This turns on a grid that helps you position a subject within the grid and autofocus. If you wish, tap the HDR button as well to increase the contrast of the image you capture.

4. Tap the Flash button in the top-left corner of the screen and tap On if your lighting is dim enough to require a flash, Off if you don't want iPhone to use a flash, or Auto if you want to let iPhone decide for you.

5. Move the camera around until you find a pleasing image. You can do a couple of things at this point to help you take your photo:

   • Tap the area of the grid where you want the camera to autofocus.

   • Pinch the screen to display a zoom control; drag the circle in the zoom bar to the right or left to crop the image to a smaller area.

6. Tap the Capture button at the bottom center of the screen. You've just taken a picture, and it's stored in the Photos app automatically.

 You can also use the Volume button with the plus sign on it (located on the left side of your iPhone) to capture a picture or start or stop video camera recording.

7. Tap the Switch Camera button in the top-right corner to switch between the front camera and rear camera. You can then take pictures of yourself, so go ahead and click the Capture button to take another picture.

8. To view the last photo taken, swipe to the left or tap the thumbnail of the latest image in the bottom-left corner of the screen; the Photos app opens and displays the photo.

9. Tap the Menu button (it's the box with an arrow coming out of it, located in the bottom-left corner of the screen) to display a menu that allows you to e-mail, or instant message the photo, assign it to a contact, use it as iPhone wallpaper, tweet it, or print it (see **Figure 16-2**).

10. To delete the image, with it displayed tap the Trash Can button in the bottom-right corner of the screen.

11. Tap the Home button to close Photos and return to the Home screen.

 To go to the camera with the Lock screen displayed, tap the Home button twice, and a camera icon appears. Tap it and you go directly to the Camera app.

 If you want to get more useful tips on getting the most out of Photos on your iPhone, then check out *iPhone Photography and Video For Dummies*, by Angelo Micheletti.

Switch Camera button

Options button

Flash button

Previously caputured image or video

Capture button

Camera/Video slider

Figure 16-1

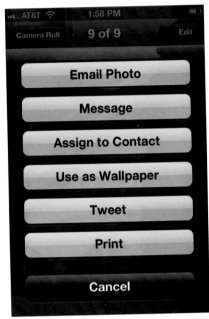

Figure 16-2

You can use the Photo Stream feature to automatically upload photos to iCloud. The photos are then downloaded to other iOS devices that are set up to download photos. Turn on Photo Stream in iPhone Settings for Photos.

Save Photos from the Web

1. The web offers a wealth of images you can download to your Photo Library. Open Safari and navigate to the web page containing the image you want.

2. Press and hold the image; a menu appears, as shown in **Figure 16-3.**

3. Tap Save Image. The image is saved to your Camera Roll folder in the Photos app, as shown in **Figure 16-4.**

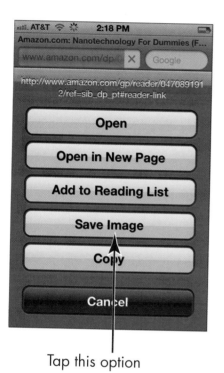

Tap this option

Figure 16-3

Figure 16-4

 For more about how to use Safari to navigate to or search for web content, see Chapter 10.

 A number of sites protect their photos from being copied by applying an invisible overlay. This blank overlay image ensures that you don't actually get the image you think you're tapping. Even if a site doesn't take these precautions, be sure that you don't save images from the Web and use them in ways that violate the rights of the person or entity that owns them.

View an Album

1. The Photos app organizes your pictures into albums, using such criteria as the folder or album on your computer from which you synced the photos or whether you captured your photo using the iPhone camera. You may also have albums for images you synced from other devices through iTunes or iCloud. To view your albums, start by tapping the Photos app icon on the Home screen.

2. If the Places tab is selected when the Photos app opens, tap the Albums tab, shown in **Figure 16-5**.

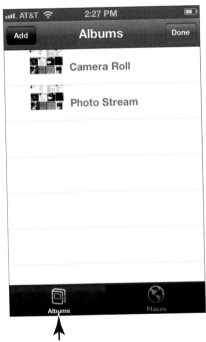

Tap this to view Albums

Figure 16-5

3. Tap an album. The photos in it are displayed.

View Individual Photos

1. Tap the Photos app icon on the Home screen.

2. Tap the Albums tab (refer to **Figure 16-5**).

3. To view a photo, tap it. The picture expands, as shown in **Figure 16-6**.

Figure 16-6

4. Flick your finger to the left or right to scroll through the album to look at the individual photos in it.

5. You can tap the Albums button (which may display the currently-open album's name) to return to the Album view.

 You can place a photo on a person's information page in Contacts. For more about how to do it, see Chapter 5.

Edit Photos

1. A new feature in iOS 5 is the ability to edit photos. Tap the Photos app on the Home screen to open it.

2. Using methods previously described in this chapter, locate a photo you want to edit.

3. Tap the Edit button; the Edit Photo screen shown in **Figure 16-7** appears.

Crop button

Red-Eye button

Enhance button

Rotate button

Figure 16-7

4. At this point, you can take four possible actions with these tools:

- *Rotate:* Tap the Rotate button to rotate the image 90 degrees at a time. Continue to tap the button to move another 90 degrees.

- *Enhance:* Tap Enhance to turn Auto-Enhance on or off. This feature optimizes the crispness of the image.

- *Red-Eye:* Tap Red-Eye if a person in a photo has that dreaded red-eye effect. When you activate this feature, simply tap each eye that needs clearing up.

- *Crop:* To crop the photo to a portion of its original area, tap the Crop button. You can then tap any corner of the image and drag inward or outward to remove areas of the photo.

5. If you're pleased with your edits, tap the Save button, and a copy of the edited photo is saved.

 Each of the four editing features has a Cancel button. If you don't like the changes you made, tap this button to stop making changes before you save the image.

Organize Photos in Camera Roll

1. If you want to create your own album, display the Camera Roll album.

2. Tap the Menu button in the top-right corner, and then tap individual photos to select them. Small check marks appear on the selected photos (see **Figure 16-8**).

3. Tap the Add To button and then tap Add to New Album. (**Note:** If you've already created albums, you can choose to add the photo to an existing album at this point.)

4. Enter a name for a new album and then tap Save. If you created a new album, it now appears in the Photos main screen with the Albums displayed.

You can also choose the Share, Copy, or Delete buttons when you've selected photos in Step 2 of this task. This allows you to share, copy, or delete multiple photos at a time.

Check marks include selected photos

Figure 16-8

Share Photos

1. You can easily share photos stored on your iPhone by sending them as e-mail attachments, via iMessage or SMS (text message), or as tweets via Twitter. First, tap the Photos app icon on the Home screen.

2. Tap the Photos tab and locate the photo you want to share.

3. Tap the photo to select it and then tap the Menu button. (It looks like a box with an arrow jumping out of it.) The menu shown in **Figure 16-9** appears.

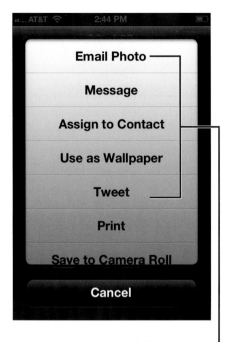

Tap one of these options

Figure 16-9

4. Tap the Email Photo, Message, or Tweet option.

5. In the message form that appears, make any modifications you want in the To, Cc/Bcc, or Subject fields and then type a message for e-mail or enter your iMessage, SMS, or tweet text.

6. Tap the Send button, and the message and photo go on their way.

 You can also copy and paste a photo into documents such as those created in the available Pages word processor application. To do this, press and hold a photo in Photos until the Copy command appears. Tap Copy and then, in the destination application, press and hold the screen and tap Paste.

Print Photos

1. If you have a wireless printer that's compatible with Apple AirPrint technology, you can print photos. With Photos open, locate the photo you want to print and tap it to maximize it.

2. Tap the Menu button and on the menu (refer to **Figure 16-9**) tap Print.

3. In the Printer Options dialog that appears (see **Figure 16-10**), tap Select Printer. iPhone searches for any compatible wireless printers on your local network.

Tap this option

Figure 16-10

4. Tap the plus or minus symbols in the Copy field to set the number of copies to print.

5. Tap the Print button, and your photo goes on its way.

Run a Slideshow

1. You can run a slideshow of your images in Photos and even play music and choose transition effects for the show. Tap the Photos app icon to open the application.

2. Display an individual photo and tap it so that an arrow appears near the bottom (this is the Slideshow button).

3. Tap the Slideshow button to see the Slideshow Options menu, shown in **Figure 16-11.**

4. If you want to play music along with the slideshow, tap the On/Off button on the Play Music field.

5. To choose music to play along with the slideshow, tap Music and, in the list that appears (see **Figure 16-12**), tap any selection from your Music library.

6. In the Slideshow Options dialog, tap Transitions and then tap the transition effect you want to use for your slideshow.

7. Tap the Start Slideshow button. The slideshow begins.

 To run a slideshow that includes only the photos contained in a particular album, tap the Albums tab, tap an album to open it, and then tap the Slideshow button to make settings and run a slideshow.

Figure 16-11

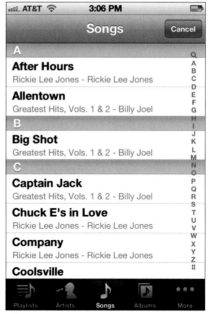

Figure 16-12

Delete Photos

1. You might find that it's time to get rid of some of those old photos of the family reunion or the last community center project. If the photos weren't transferred from your computer but instead were downloaded or captured as screenshots on the iPhone, you can delete them. Tap the Photos app icon on the Home screen.

2. Tap the Albums or Photos tab and then tap an album to open it.

3. Locate a photo you want to delete, and then tap it to display the buttons. Tap the Trash Can icon. In the confirming dialog that appears (shown in **Figure 16-13**), tap the Delete Photo button to finish the deletion.

Tap this option

Figure 16-13

Getting the Most Out of Video Features

Two preinstalled applications on your iPhone exist to help you view videos. Using the Videos app, you can watch downloaded movies or TV shows or media you've synced from iCloud or your Mac or PC. The YouTube app takes you online to the popular video-sharing site. Videos there range from professional music videos to clips from news or entertainment shows and personal videos of cats dancing (and other news-making events). If you like to view things on a bigger screen, you can even use iPhone's AirPlay to send your iPhone movies and photos to your TV wirelessly if your TV is equipped with Apple TV ($99).

In addition, iPhone sports both a front and rear video camera you can use to capture your own videos, and by purchasing the iMovie app for $4.99 (a more limited version of the long-time mainstay on Mac computers), you add the capability to edit those videos.

In this chapter, I explain all about shooting, watching, and editing video content from a variety of sources. For practice, you might want to refer to Chapter 12 first to purchase or download one of many available TV shows or movies.

Capture Your Own Videos with the Built-in Cameras

1. To capture a video, tap the Camera app on the Home screen. In iPhone, two video cameras are available to capture video from either the front or back of the device, making it possible for you to take videos that you can then share with others or edit. (See more about this topic in the next task.)

2. The Camera app opens. Use the Camera/Video slider to switch from the still camera to the video camera (see **Figure 17-1**).

Flash button   Switch Camera button

Last media captured | Camera/Video slider
Record button

Figure 17-1

3. If you want to switch between the front and back cameras, tap the Switch Camera button in the top-right corner of the screen (refer to **Figure 17-1**).

4. Tap the Record button to begin recording the video. (This button flashes when the camera is recording.) When you're finished, tap the Record button again. Your new video is now listed in the bottom-left corner of the screen.

 Before you start recording, make sure you know where the camera lens is — while holding the iPhone and panning, you can easily put your fingers directly over the lens!

Edit Movies with the iMovie App

1. A version of the iMovie app is available for the iPhone. After you capture a video, you can use iMovie to edit it. If you want to edit the video you captured in the previous task, start by purchasing iMovie from the App Store (it costs $4.99).

2. Tap the iMovie app icon on the Home screen to open it.

3. On the screen that appears, tap the New Project button (see **Figure 17-2**) to begin a new movie project. Tap the Insert button, and any videos you've taken (see the previous task) are listed at the top of the screen; double-tap one to open it (see **Figure 17-3**).

4. To scroll through your video, flick your finger on the storyboard. Wherever the red line sits (refer to **Figure 17-3**) is where your next action, such as playing the video or adding a transition, begins.

5. To add one video to the end of another, double-tap another clip in the list of media. The clip appears to the right of the first one in the storyboard.

New Project button

Figure 17-2

6. If you want to add music to your videos, tap the Settings button (refer to **Figure 17-3**) and tap to turn Theme Music on, and if you wish the music to repeat tap Loop Background music (a quick and easy way to add a musical track to your video).

7. To record an audio narration, scroll to the point in the movie where you want it to be heard and tap the Microphone button (refer to **Figure 17-3**).

8. Tap the Record button, tap Record, wait for the countdown to finish, and then record. When you're done recording, tap Stop (see **Figure 17-4**).

9. In the options shown in **Figure 17-5**, tap Review to hear your recording, Accept to save the recording, or Discard or Retake if you're not completely happy with it. If you accept the recording, it appears below the storyboard at the point in the movie where it will play back.

Media preview

Microphone button | Settings button

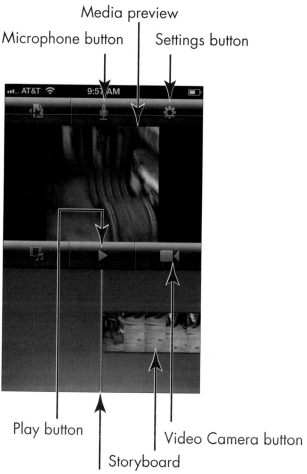

Play button

Storyboard

Video Camera button

Video progression indicated by red line

Figure 17-3

You can do more things with iMovie, such as add transitions between clips or rearrange and cut segments out of clips. Play around with iMovie to see the possibilities, or check out *iLife '11 For Dummies,* by Tony Bove, and *iPhone Photography & Video For Dummies,* by Angelo Micheletti.

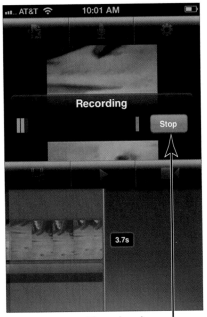

Tap this button

Figure 17-4

Figure 17-5

 To quickly return from iMovie to your camera, perhaps to capture more footage, tap the Video Camera button (refer to **Figure 17-3**).

 You can tap the Settings button (refer to **Figure 17-3**) to open a menu that lets you make settings for special effects that allow you to to fade your movie in or out at the beginning and end.

Play Movies, Podcasts, or TV Shows with Videos

1. Tap the Videos app icon on the Home screen to open the application.

2. On a screen like the one shown in **Figure 17-6,** tap the program you want to watch.

Information about the program appears, as shown in **Figure 17-7.**

Tap the program you want to watch

Figure 17-6

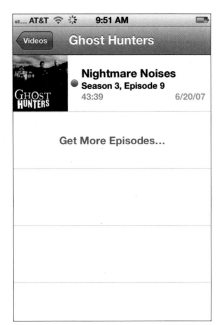

Figure 17-7

3. Tap the item again, and it loads; tap the Play button, and the movie, TV show, or podcast begins playing. Note that the progress of the playback is displayed on the Progress bar (see **Figure 17-8**), showing how many minutes you've viewed and how many remain. If you don't see the bar, tap the screen once to display it briefly, along with a set of playback tools at the bottom of the screen.

4. With the playback tools displayed, take any of these actions:

- Tap the Pause button to pause playback.

- Tap either Go to Previous Chapter or Go to Next Chapter to move to a different location in the video playback.

- Tap the circular button on the Volume slider and drag the button left or right to decrease or increase the volume, respectively.

Done button
Progress bar
Pause/Play button

Go to Previous Chapter button | Go to Next Chapter button
Volume slider

Figure 17-8

5. To stop the video and return to the information screen, tap the Done button to the left of the Progress bar.

Note that if you've watched a video and stopped it partway, it opens by default to the last spot you were viewing. To start a video from the beginning, tap and drag the circular button on the Progress bar all the way to the left.

If your controls disappear during playback, just tap near the top or bottom of the screen, and they'll reappear.

Turn on Closed-Captioning

1. iTunes and iPhone offer support for closed-captioning and subtitles. If a movie you purchased or rented has either closed-captioning or subtitles, you can turn on the feature in iPhone. Look for the CC logo on media you download to use this feature; be aware that video you record won't have this capability. Begin by tapping the Settings icon on the Home screen.

2. On the screen that appears, tap Video in the Settings section on the left side of the screen.

3. On the menu that displays (see **Figure 17-9**), tap the Closed Captioning On/Off button to turn on the feature. Now when you play a movie with closed-captioning, you can click the Audio and Subtitles button to the left of the playback controls to manage these features.

Make sure this is set to On

Figure 17-9

Delete a Video from the iPhone

You can buy videos directly from your iPhone, or you can sync via iCloud or iTunes to place content you've bought on another device on your iPhone.

When you want to get rid of video content on your iPhone because it's a memory hog, you can delete it using iTunes and then sync again, removing it from your iPhone. If you buy a video using iTunes, sync to download it to your iPhone, and then delete it from your iPhone, it's still saved in your iTunes library. You can sync your computer and iPhone again to download the video once more. Remember, however, that rented movies, once deleted, are gone with the wind.

 iPhone has much smaller storage capacity than your typical computer, so downloading lots of TV shows or movies can fill its storage area quickly. If you don't want to view an item again, delete it to free up space.

Find Videos on YouTube

1. Although you *can* go to YouTube and make use of all its features using iPhone's Safari browser, there's an easier way: using the dedicated YouTube application that's pre-installed on your iPhone. This app provides buttons you can tap to display different content and features using your touchscreen. Tap the YouTube icon on the Home screen to open the app.

2. Tap the Featured button at the bottom of the screen, if it's not already selected (see **Figure 17-10**).

3. To find videos, tap the Search button at the bottom of the screen; a search field appears. Tap in the field, and the keyboard opens.

4. Type a search term and tap the Search key on the keyboard.

5. Use your finger to scroll down the screen to see additional results.

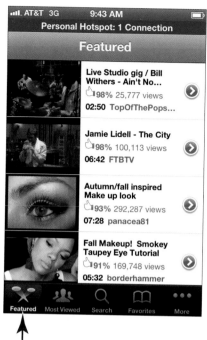

Tap this button

Figure 17-10

6. To display the top-rated or most-viewed videos, tap the Most Viewed button at the bottom of the screen or tap the More button and choose Most Recent or Top Rated categories. You can also look at the history of videos you've viewed, access videos you've saved to the YouTube My Videos feature, or check out subscriptions you've taken out or playlists you've created in YouTube.

7. When you find a movie you want to view, tap it to display it. The movie begins loading; see the next task for details about how to control the playback.

> When you load a movie, you can use the Related and More From tabs in the playback window to find additional content related to the topic, or find more videos posted on YouTube by the same source.

If you find a movie you like, you might like to add it to a YouTube playlist. Tap Done when the video is over and then tap the back arrow with the name of the video on it to return to general information about the video. There you'll find Add to Playlist and Share buttons. Tap Playlist to sign into your YouTube account and add the video to a playlist.

Control Video Playback

1. With a video displayed (see the previous task), tap the Play button. The video plays (see **Figure 17-11**).

Full Screen button
Playback progress bar

Play/Pause button

Figure 17-11

2. Tap the Pause button to pause playback. (If the button isn't visible, tap the screen once to display it.)

3. To move forward or backward in the movie, tap the circular button on the Playback progress bar and drag the button to the right or left.

When the video is finished, you can replay it from the beginning by tapping the Play button again.

Change Widescreen Media Views

1. In the previous task, you watched a video in the standard view. To enlarge the video for the full screen, tap the Full Screen button (refer to **Figure 17-11**) and the video displays in a full-screen version with a black border.

2. To return the movie to its smallest size, tap the Full Screen button again.

You can use the double-tap method of moving between the two screen formats. Double-tapping the smaller of the two moves you to the largest full-screen view. Double-tapping the largest full-screen view zooms back to the smaller view.

Flag Content as Inappropriate

1. If there's a chance your grandkids can get hold of your iPhone, you might appreciate the capability to flag inappropriate content on YouTube. First you have to set a restriction in your YouTube account using your computer and then set a flag using the iPhone YouTube app, which causes a passcode to be required in order to access that content. Locate a video in a category such as Featured or Favorites and tap the arrow button to the right of it, as shown in **Figure 17-12**.

2. In the screen that appears, tap the arrow to the right of the video title.

3. Scroll down the More Info screen that appears and tap the Rate, Comment, or Flag button shown in **Figure 17-13**.

Tap this button   Tap this button

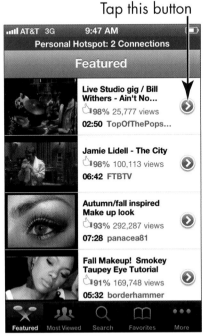

Figure 17-12   **Figure 17-13**

4. Tap the Flag as Inappropriate option.

5. Sign into your YouTube account to complete the action.

Rate Videos

1. Locate a video in a category such as Featured or Favorites and tap the arrow button to the right of it (refer to **Figure 17-12**).

2. In the screen that appears, tap the arrow to the right of the video title.

3. Scroll down the More Info screen that appears and tap the Rate, Comment, or Flag button (refer to **Figure 17-13**) and then tap Rate or Comment.

4. In the screen that appears (see **Figure 17-14**) tap the Like or Dislike button and then tap the Send button.

Tap Like or Dislike...   then tap Send

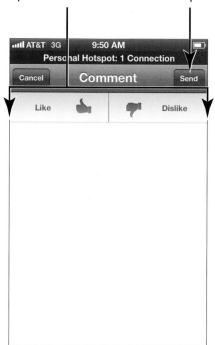

Figure 17-14

5. You're asked to sign in to your YouTube account. After you sign in, your Like or Dislike rating is accepted.

 You can view the highest-rated videos on YouTube by tapping the More button at the bottom of the screen and then tapping Top Rated.

Share Videos

1. Find a video you want to share in a category such as Favorites or Features and tap the arrow to the right of it.

2. In the information screen shown in **Figure 17-15,** tap the Share Video button. Tap either Mail Link to This Video or Tweet.

3. In the e-mail form shown in **Figure 17-16,** enter a recipient in the To field and add to the message, if you like. If you chose to post a tweet, enter your message in the Tweet form.

Tap this button

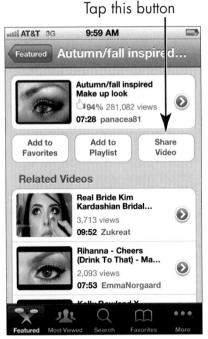

Figure 17-15

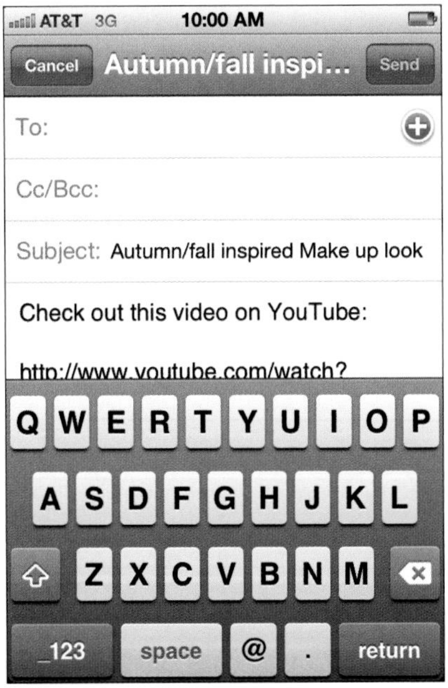

Figure 17-16

4. Tap the Send button to send someone a link to the video or post your tweet.

Add to Video Favorites

1. Locate a video in a category such as Favorites or Featured and tap the button to the right of the video.

2. Tap the Add to Favorites button (refer to **Figure 17-15**). The video has been added to your Favorites.

3. To view your favorite movies, tap the Favorites button at the bottom of the YouTube screen. Your favorite videos are displayed.

 To delete a favorite while you're on the Favorites screen, tap the Edit button. Delete buttons appear on each movie. Tap the movie-specific Delete button to remove that movie. Tap the Done button to leave Editing mode.

Playing Games

The iPhone is super for playing games, with its bright screen, portable size, and ability to rotate the screen as you play and track your motions. You can download game apps from the App Store and play them on your device. You can also use the preinstalled Game Center app to help you find and buy games, add friends to play against, and track scores.

In this chapter, you get an overview of game playing on your iPhone, including opening a Game Center account, adding a friend, purchasing and downloading games, and playing basic games solo or against friends.

Chapter 18

Get ready to . . .

Of course, you can also download games from the App Store and play them on your iPhone without having to use Game Center. What Game Center provides is a place where you can create a gaming profile, add a list of gaming friends, keep track of your scores and perks, and shop for games (and only games) in the App Store, which includes listings of top-rated games and game categories to choose from.

Open an Account in Game Center

1. Using the Game Center app, you can search for and buy games, add friends with whom you can play those games, and keep records of your scores for posterity. From the Home screen, tap the Game Center icon. If you've never used Game Center, you're asked whether to allow *push notifications:* If you want to receive these notices alerting you that your friends want to play a game with you, tap OK. You should, however, be aware that push notifications can drain your iPhone's battery.

2. On the Game Center opening screen (see **Figure 18-1**), tap Create New Account.

3. If the correct country isn't listed in the New Account dialog, tap in the Location field and select another location. If the correct location is already showing, tap Next to confirm it.

4. In the next dialog you see, tap the Month, Day, and Year fields, enter your date of birth, and then tap Next.

5. In the Game Center Terms & Conditions dialog, swipe to scroll down (and read) the conditions, and then tap Agree if you want to continue creating your account. A confirmation dialog appears; tap Agree once more to accept for real!

Tap here

Figure 18-1

6. In the next dialog that opens, tap each field and enter your name, e-mail address, and password information. Tap the Next button above the onscreen keyboard to move to the next field.

7. Tap the Question field to select a security question to identify yourself, and tap the Answer field and type in an answer to the question. Be sure to scroll to the bottom of this dialog and choose to turn off the e-mail notification subscription if you don't want to have Game Center send you messages. Tap Next to proceed, and then sign into your account. See the following task, "Create a Profile," to create your Game Center profile in subsequent dialogs.

 When you first register for Game Center, if you use an e-mail address other than the one associated with your Apple ID, you may have to create a new Apple

ID and verify it using an e-mail message that's sent to your e-mail address. See Chapter 3 for more about creating an Apple ID when opening an iTunes account.

Create a Profile

1. When you reach the last dialog in Step 7 of the previous task, you're ready to create your profile and specify some account settings. You can also make most of these settings after you've created your account, by tapping your Account name on the Game Center home screen and then tapping View Account. In the Create Profile dialog (Account dialog for an existing account) that appears (see **Figure 18-2**), if you don't want other players to be able to invite you to play games when Game Center is open, tap the Game Invites On/Off button to turn off the feature.

2. If you don't want other players to be able to see your real name, tap the Public Profile On/Off button to turn this feature off.

3. If you want your friends to be able to send you requests for playing games via e-mail, check to see if the e-mail address listed in this dialog is the one you want them to use. If not, tap Add Another Email and enter another e-mail address.

4. In the Nickname field, enter the handle you want to be known by when playing games.

5. Tap Done when you're finished with the settings. You return to the Game Center home screen, already signed in to your account with information displayed about friends, games, and gaming achievements (all at zero initially.)

6. To add a picture to your profile, tap Change Photo.

7. The two options of Take Photo or Choose Photo appear. Tap Choose Photo to select a photo from your camera roll or a photo library. Tap the library you want to use, and scroll to locate the photo.

8. Tap the photo, and it appears in a Choose Photo dialog. You can use your finger to move the photo around or scale it, and then tap Use. The photo now appears on your Game Center home screen (see **Figure 18-3**).

Your photo appears here

Figure 18-2

Figure 18-3

 After you create an account and a profile, whenever you go to the Game Center, you log in by entering your e-mail address and password and then tapping Sign In.

 You can change account settings from the Game Center home screen: Tap your account name and then View Account, make changes to your settings, and then tap Done.

Add Friends

1. If you want to play Game Center games with others who have an Apple ID and an iPad, iPod touch, or iPhone, add them as friends so that you can invite them to play. From the Game Center home screen, tap the Friends button at the bottom of the screen.

2. On the Friends page, tap Add Friends.

3. Enter an e-mail address in the To field and edit the invitation, if you like.

4. Tap the Send button. After your friend accepts your invitation, his or her name is listed on the Friends screen.

 With iOS 5, Game Center gained a Friend Recommendations feature. Tap the Friends tab, and then tap the A-Z button in the top-left corner. A Recommendations section appears above the list of your current friends. These are people who play the same or similar games, so if you like, try adding one or two as friends.

 You will probably also receive requests from friends who know you're on Game Center. When you get one of these e-mail invitations, be sure that you know the person sending it before you accept; otherwise, you could be putting a stranger in communication with yourself.

Purchase and Download Games

1. Time to get some games to play! Open Game Center and sign in to your account.

2. Tap the Games button at the bottom of the screen, scroll to the bottom, and then tap Find Game Center Games (see **Figure** 18-4).

Tap here to find more games

Figure 18-4

3. In the list of games that appears, scroll through the list of featured games. To view different games, tap either the Top 25 or Categories button at the bottom of the screen. *Note:* Accessing these from the Game Center displays only game apps, as opposed to accessing apps from the App Store, which shows you all categories of apps.

4. To search for a particular title, tap the Search button and enter the name by using the onscreen keyboard.

5. Tap a game title to view information about it. To buy a game, tap the button labeled with either the word *free* or the price (such as $1.99). Then tap the button again, which is now labeled Buy Now.

6. A dialog appears, asking for your Apple ID and password. Enter these and tap OK.

7. Another verification dialog appears, asking you to sign in. Follow the instructions on the next couple of screens to enter your password and verify your payment information if this is the first time you've signed in to your account from this device.

8. When the verification dialog appears, tap Buy. The game downloads.

 If you've added friends to your account, you can go to the Friends page and view games that your friends have downloaded. To purchase one of these games, just tap it in your friend's list and, at the top of the screen that appears, tap the button labeled with the game's price.

 You may see buttons labeled Buy It Now or Available at the App Store while you're exploring game recommendations in the Games section of Game Center. Tapping such a button takes you from Game Center directly to the App Store to buy the game.

Master iPhone Game-Playing Basics

It's almost time to start playing games, but first let me give you an idea of iPhone's gaming strengths. For many reasons, iPhone may be the ultimate gaming device because of the following strengths:

➡ **Fantastic-looking screen:** First, the iPhone screen offers bright colors for great gaming. It's no lie: See the *Angry Birds* game in **Figure 18-5,** for example. In-Plane Switching (IPS) technology lets you hold your iPhone at almost any angle (it has a 178-degree viewing angle) and still see good color and contrast.

➡ **Faster processor:** The super-fast A5 processor chip in your iPhone 4S is ideal for gaming.

Figure 18-5

⟶ **Long battery life:** A device's long battery life means that you can tap energy from it for many hours of gaming fun.

⟶ **Specialized game-playing features:** Some newer games have features that take full advantage of the iPhone's capabilities. For example, *Nova* (from Gameloft) features Multiple Target Acquisition, which lets you target multiple bad guys in a single move to blow them out of the water with one shot. In *Real Racing Game* (Firemint), for example, you can look in your rearview mirror as you're racing to see what's coming up behind you, a feature made possible by the iPhone's larger screen.

⟶ **Great sound:** The built-in iPhone speaker is a powerful little item, but if you want an experience that's even more up close and personal, you can plug in a headphone, some speaker systems, or microphone using the built-in jack.

 The iPhone has a built-in motion sensor — the *three-axis accelerometer* — as well as a gyroscope. These features provide lots of fun for people developing apps for the iPhone, as they use the automatically rotating screen to become part of the gaming experience. For example, a built-in compass device reorients itself automatically as you switch your iPhone from land-scape to portrait orientation. In some racing games, you can grab the iPhone as though it were a steering wheel and rotate the device to simulate the driving experience.

Play against Yourself

Many games allow you to play a game all on your own. Each has dif-ferent rules and goals, so you'll have to study a game's instructions and help to learn how to play it, but here's some general information about these types of games:

➠ Often a game can be played in two modes: with oth-ers or in a *solitaire version*, where you play yourself or the computer.

➠ Many games you may be familiar with in the offline world, such as Carcassonne or Scrabble, have online versions. For these, you already know the rules of play, so you simply need to figure out the execution. For example, in the online Carcassonne solitaire game, you tap to place a tile on the board, tap the placed tile to rotate it, and tap the check mark to complete your turn and reveal another tile.

➠ All the games you play on your own record your scores in Game Center so you can see how you're progressing in building your skills.

Play Games with Friends in Game Center

1. After you have added a friend and both of you have downloaded the same games, you're ready to play. The rules of play are different for each game, but here are the basic steps for getting a game going. Tap the Game Center app icon on the Home screen and sign in, if necessary.

2. Tap Friends. The Friend page (see **Figure 18-6**) appears.

Figure 18-6

3. Tap the name of the friend you want to play and then tap the name of a game you have in common. At this point, some games offer you an invitation to send to the friend — if so, wait for your friend to respond, which he does by tapping Accept or Decline on his device.

4. The game should appear, and you can tap Play to start playing according to whatever rules the game has. Your scores mount up as you play.

5. When you're done playing, tap either Friend or Game in Game Center to see your score and your friend's score listed. Game Center tracks your achievements, including points and perks that you've earned along the way. You can also compare your gaming achievements to those of top-ranking players across the Internet — and check your friends' scores by displaying the Friends page and then tapping the Points tab near the top of the screen (refer to **Figure 18-6**).

 If your friends aren't available, you can play a game by tapping its title on the Games page and then tapping Play. You can then compare your scores with others around the world who have also played the game recently.

Finding Your Way with Maps

*I*f you're new to the Maps app, you'll find it has lots of useful functions. You can find directions with suggested alternate routes from one location to another. You can bookmark locations to return to them again. And, the Maps app makes it possible to get information about locations, such as the phone numbers and web links to businesses. You can even look at a building or location as though you were standing in front of it on the street, add a location to your Contacts list, or e-mail a location link to your buddy.

Be prepared: This application is seriously cool, and you're about to have lots of fun exploring it in this chapter.

Go to Your Current Location

1. iPhone can figure out where you are at any time and display your current location. From the Home screen, tap the Maps icon. Tap the Current Location button (the small, blue arrow to the left of the Search button; see **Figure 19-1**).

Settings button

Current Location button

Figure 19-1

2. A map is displayed with a blue pin indicating your current location and a blue circle around it (refer to **Figure 19-1**). Depending on your connection, Wi-Fi or 3G, the circle indicates the area surrounding your location based on cell tower triangulation — your exact location can be anywhere within the area of the circle, and it's likely to be less accurate using a Wi-Fi connection than with your phone's 3G connection.

3. Double-tap the screen to zoom in on your location. (Additional methods of zooming in and out are covered in the "Zoom In and Out" task, later in this chapter.)

 If you access maps via a Wi-Fi connection, your current location is a rough estimate based on a triangulation method. By using your 3G data connection, you access the global positioning system (GPS), which can more accurately pinpoint where you are. Still, if you type a starting location and an ending location to get directions, you can get pretty accurate results even with a Wi-Fi–connected iPhone.

Change Views

1. The Maps app offers four views: Standard, Satellite, Hybrid, and List. iPhone displays the Standard view (see the top-left image in **Figure 19-2**) by default the first time you open Maps. To change views, with Maps open, tap the Settings button in the bottom-right corner of the screen (refer to Figure 19-1) to turn the page, so to speak, and reveal the Maps menu, shown in **Figure 19-3**.

2. Tap the Satellite option. The Satellite view (refer to the top-right image in **Figure 19-2**) appears.

3. Tap the Settings button to reveal the menu again, and then tap Hybrid. In Hybrid view, Satellite view is displayed with street names superimposed (refer to the bottom-left image in **Figure 19-2**).

4. Finally, tap the Settings button to reveal the menu one more time, and then tap List. A list of locations on the map is displayed (refer to the bottom-right image in **Figure 19-2**).

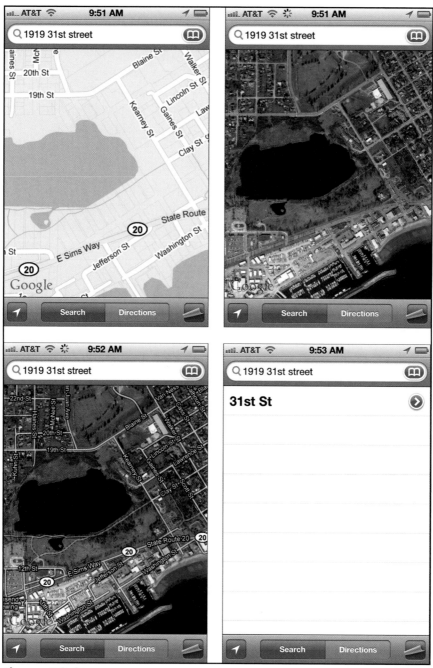

Figure 19-2

Figure 19-3

 On the Maps menu, you can also access a Traffic overlay feature. If you live in a larger metropolitan area (this feature doesn't really work in small towns), turn on this feature by tapping the Show Traffic button. The traffic overlay shows different colors on roads indicating accidents or road closures to help you navigate your rush hour commute or trip to the mall. Green tells you you're good to go, yellow indicates slowdowns, and red means avoid at all costs.

 You can drop a pin to mark a location on a map that you can return to. See the task "Drop a Pin," later in this chapter, for more about this topic.

 To print any displayed map to an AirPrint-compatible wireless printer, just tap the Print button on the Maps menu.

Zoom In and Out

1. You'll appreciate the Zoom feature because it gives you the capability to zoom in and out to see more or less detailed maps and to move around a displayed map. With a map displayed, double-tap with a single finger to zoom in (see **Figure 19-4**).

Figure 19-4

2. Double-tap with two fingers to zoom out, revealing less detail.

3. Place two fingers positioned together on the screen and move them apart to zoom in.

4. Place two fingers apart on the screen and then pinch them together to zoom out.

5. Press your finger to the screen and drag the map in any direction to move to an adjacent area.

It can take a few moments for the map to redraw itself when you enlarge, reduce, or move around it, so try a little patience. Areas that are being redrawn look like blank grids but are filled in eventually. Also, if you're in Satellite or Hybrid view, zooming in may take some time; wait it out because the blurred image resolves itself.

Go to another Location

1. With Maps open, tap in the Search field (see **Figure 19-5**); the keyboard opens.

2. Type a location, using a street address with city and state, or a destination such as *Empire State Building* or *Detroit airport*. Maps may make suggestions as you type if it finds any logical matches. Tap the Search key on the keyboard, and the location appears with a red pin inserted in it and an information bar with the location, an Information icon, and in some cases the Street view icon (see **Figure 19-6**). Note that if several locations match your search term, several pins may be displayed.

 Try asking for a type of business or location by zip code. For example, if you crave something with pepperoni, enter **99208 pizza**.

3. You can also tap the screen and drag in any direction to move to a nearby location.

4. Tap the Bookmark icon (the little book symbol to the right of the Search field; refer to **Figure 19-6**), and then tap the Recents tab to reveal recently visited sites. Tap a bookmark to go there.

 As you discover later in this chapter, in the "Add and View a Bookmark" task, you can also quickly go to any location you've previously visited and saved using the Bookmarks feature.

Street view icon

Information bar

Bookmark icon

Figure 19-5

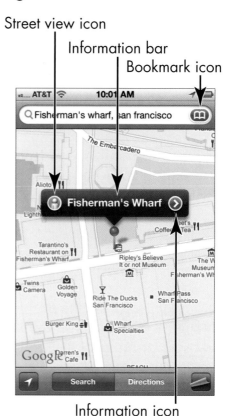

Information icon

Figure 19-6

 If you enter a destination such as *Bronx Zoo,* you might want to also enter its city and state. Entering *Bronx Zoo* landed me in the Woodland Park Zoo in Tacoma because Maps looks for the closest match to your geographical location in a given category.

Drop a Pin

1. Pins are markers: A green pin marks a start location, a red pin marks a search result, and a blue pin (referred to as the *blue marker*) marks your iPhone's current location. If you drop a pin yourself, it appears in a lovely purple. Display a map that contains a spot where you want to drop a pin to help you find directions to or from that site.

2. If you need to, you can zoom in to a more detailed map to see a better view of the location you want to pin.

3. Press and hold your finger on the screen at the location where you want to place the pin. The pin appears, together with an information bar (refer to **Figure 19-6**).

4. Tap the Information icon (refer to **Figure 19-6**) on the information bar to display details about the pin location (see **Figure 19-7**).

Figure 19-7

 To delete a pin you've dropped, tap the pin to display the information bar and then tap the Information icon. In the Information dialog that opens, tap Remove Pin. This method works only with pinned sites that aren't bookmarked.

Add and View a Bookmark

1. A *bookmark* provides a way to save a destination so you can display a map or directions to it quickly. To add a bookmark to a location, first place a pin on it, as described in the preceding task.

2. Tap the Information icon to display the Information dialog.

3. Tap the Add to Bookmarks button.

4. The Add Bookmark dialog (see **Figure 19-8**) and the keyboard appear. If you like, you can modify the name of the bookmark.

Modify the bookmark name here

Figure 19-8

5. Tap Save.

6. To view your bookmarks, tap the Bookmark icon (it looks like a little open book; refer to **Figure 19-6**) at the top of the Maps screen. Be sure that the Bookmarks tab is selected; a list of bookmarks is displayed.

7. Tap on a bookmark to go to the location.

 You can also view recently viewed locations, even if you haven't bookmarked them. Tap the Bookmark icon and then, at the bottom of the Bookmarks dialog that appears, tap Recents. Locations you've visited recently are listed there. Tap one to return to it.

Delete a Bookmark

1. Tap the Bookmark icon and then tap the Bookmarks tab at the bottom of the dialog that appears, to be sure you're viewing Bookmarks.

2. Tap the Edit button. A red minus icon appears to the left of your bookmarks, as shown in **Figure 19-9**.

3. Tap a red minus icon.

4. Tap Delete. The bookmark is removed.

 You can also use a touchscreen gesture after you've displayed the Bookmarks in Step 1 in the preceding list. Simply swipe across a bookmark and then tap the Delete button.

 You can clear out all recent locations stored by Maps to give yourself a clean slate. Tap the Bookmark icon and then tap the Recents tab. Tap Clear and then confirm by tapping Clear All Recents.

Red minus icons

Figure 19-9

Find Directions

1. You can get directions in a couple of different ways. With at least one pin on your map in addition to your current location, tap the Directions tab and then tap Route. A line appears, showing the route between your current location and the closest pin for the currently selected transportation method: by car, on foot, or by using public transit (see **Figure 19-10**).

2. To show directions from your current location to another pin, tap the other pin, and the route is redrawn.

3. You can also enter two locations to get directions from one to the other. With the Directions tab selected in Maps, tap the Edit button (see **Figure 19-11**) and then tap in the field labeled *Current Location*. The keyboard appears.

Tap this button

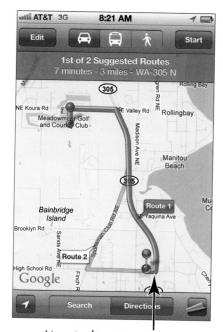

Line indicating your route

Figure 19-10

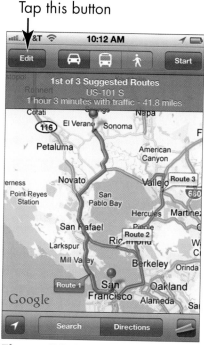

Figure 19-11

4. Enter a different starting location.

5. Tap in the Destination field, enter a destination location, and then press the Search button on the keyboard. The route between the two locations is displayed.

6. You can also tap the Information icon on the information bar that appears above any selected pin and use the Directions to Here or Directions from Here button to generate directions (refer to **Figure 19-7**).

7. When a route is displayed, a blue bar appears along the top of the Maps screen (refer to **Figure 19-10**) with information about the distance and time it takes to travel between the two locations. Here's what you can do with this informational display:

- Tap the car, bus, or pedestrian icon to get driving, public transportation, or walking directions.

- Tap Start to display step-by-step directions (see **Figure 19-12**); the arrow keys in the top-right corner take you through the directions one step at a time.

8. If there are alternate routes, Maps notes the number of alternate routes in the informational display and shows the routes on the map. Tap a route number to make it the active route (see **Figure 19-13**).

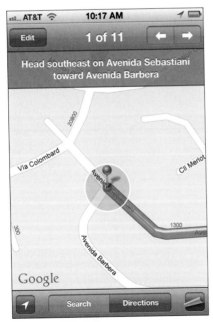

Figure 19-12

Tap a route number to display it

Figure 19-13

In Directions view in Maps, after you generate directions from one location to another, tap the Edit button (refer to **Figure 19-11**). Notice the button with a zigzag line between the Current Location and Destination fields; after you generate directions from one location to another, tap this button to generate reverse directions. Believe me: They aren't always the same — especially when one-way streets are involved!

View Information about a Location

1. You've displayed the Information dialog for locations to add a bookmark or get directions in previous tasks. Now you focus on the useful information displayed there. Go to a location and tap the pin.

2. On the information bar that appears above the pinned location, tap the Information icon (refer to **Figure 19-6**).

3. In the Information dialog (refer to **Figure 19-7**), tap the web address listed in the Home Page field, which you can use to go the location's web page, if it has one associated with it.

4. You can also press and hold either the Phone or Address field and use the Copy button to copy the phone number, for example, so that you can place it in a Notes document for future reference.

5. Tap outside the Information dialog to close it.

Rather than copy and paste information, you can easily save all information about a location in your Contacts address book. See the "Add a Location to a Contact" task, later in this chapter, to find out how it's done.

View a Location from Street Level

1. You can view only certain locations from street level, so you have to explore to try out this feature. On the Search tab of Maps, tap in the Search field and enter a location, such as your favorite local shopping mall.

2. When the location appears, tap the Street view icon on its information bar. Street view appears.

3. When you're in Street view (see **Figure 19-14**), you can tap and drag the screen to look around you in all directions. Remember the view is a moment frozen in time, perhaps from a few years ago.

Tap here to return to the standard map view
Figure 19-14

4. Tap the small circular map in the bottom-right corner (refer to **Figure 19-14**) to return to the standard map view.

 You can also drag the screen down to get a better look at tall skyscrapers or drag up to view the street and its manhole covers. The small, circular map in the bottom-right corner highlights what you're looking at in the specific moment. In addition, street names are displayed down the center of streets.

Use the Compass

1. The Compass feature works from only your current location, so start by tapping the Current Location icon at the bottom left corner of the Maps screen (refer to **Figure 19-1**).

2. Tap the Current Location icon again to turn on the Compass. The icon changes to the shape of a small beacon (see **Figure 19-15**).

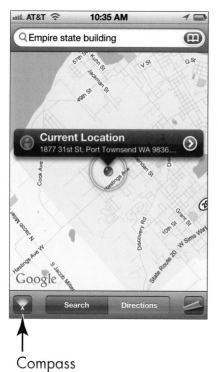

Compass

Figure 19-15

3. Move your iPhone around in different directions and note that the map moves as well, indicating which direction you're facing.

4. To turn off the Compass, tap the Current Location icon one more time.

A message may appear indicating that the Compass needs resetting because of interference. To do so, move away from any electronic equipment that might be causing the problem and move the iPhone around in what Apple describes as a figure-eight motion.

 Location Services has to be turned on in iPhone Settings for the Compass feature to be available.

Add a Location to a New Contact

1. Tap a pin to display the information bar.

2. Tap the Information icon.

3. In the Information dialog that appears, tap Add to Contacts.

4. In the resulting dialog, tap Create New Contact. The New Contact dialog appears (see **Figure 19-16**). (Note if you choose Add to an Existing Contact here you can add choose a contact from your Contacts list to add the location to).

5. Whatever information was available about the location has already been entered. Enter any additional information you need, such as a name, phone number, or e-mail address.

6. Tap Done. The information is stored in your Contacts address book.

 You can choose a distinct ringtone or text tone for a new contact. Just tap the Ringtone or Text Tone field in the New Contact form to see a list of options. When that person calls either on the phone or via FaceTime, or texts you via SMS or iMessage, you will recognize him or her from the type of tone that plays.

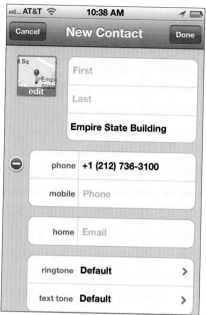

Figure 19-16

Share Location Information

1. Tap a pin to display the information bar.

2. Tap the Information icon.

3. In the Information dialog that appears, tap Share Location. In the dialog that appears, you can choose to share via text message, Tweet, or e-mail. Tap Email to see how this option works.

4. On the e-mail form that appears (see **Figure 19-17**), use the onscreen keyboard to enter a recipient's e-mail address and any Cc/Bcc addresses, and add or change the subject or message as you like.

Figure 19-17

5. Tap Send. A link to the location information in Google Maps is sent to your designated recipients.

If you choose Tweet in Step 3, you have to have installed the Twitter app and have a Twitter account set up using iPhone Settings. Tapping Message in Step 3 displays a new message form; just enter an e-mail address or phone number in the To: field, enter your text message, and then tap Send.

Part V

Managing Your Life and Your iPhone

The 5th Wave   By Rich Tennant

"Other than this little glitch with the landscape view, I really love my iPhone."

Keeping On Schedule with Calendar

Whether you're retired or still working, you have a busy life full of activities (even busier if you're retired, for some unfathomable reason). You may need a way to keep on top of all those activities and appointments. The Calendar app on your iPhone is a simple, elegant, electronic daybook that helps you do just that.

In addition to being able to enter events and view them in a list or by the day or month, you can set up Calendar to send alerts to remind you of your obligations and search for events by keywords. You can even set up repeating events, such as weekly poker games, monthly get-togethers with the girls or guys, or weekly babysitting appointments with your grandchild. To help you coordinate calendars on multiple devices, you can also sync events with other calendar accounts.

In this chapter, you master the simple procedures for getting around your calendar, entering and editing events, setting up alerts, syncing, and searching.

View Your Calendar

1. Calendar offers several ways to view your schedule. Start by tapping the Calendar app icon on the Home screen to open it.

2. Tap the List button at the bottom of the screen to display List view (if it's not already displayed). This view, shown in **Figure 20-1**, displays your daily appointments for every day in a list with times listed on the left. Tap an event in the list to get more event details, or tap the + symbol to add an event.

 If you'd like to display events only from a particular calendar, such as the Birthday or US Holidays calendars, tap the Calendars button in the top-left corner of the List view and select a calendar to base the list on.

3. Tap the Day button to view all events for the current day, as shown in **Figure 20-2**. In this view, appointments appear like a page in a daily appointment book. (Note that in landscape orientation, you see the day as a column in a set of columns for the days of the week.)

4. Tap the Month button to get an overview of your busy month (see **Figure 20-3**). In this view, you see the calendar for the month with any events for the selected day displayed in a list at the bottom of the view. Again, in landscape orientation you will instead see the selected day of the month in a series of daily columns.

5. To move from one month to the next in Month view, tap the Next or Previous buttons to the right and left of the month name at the top of the calendar.

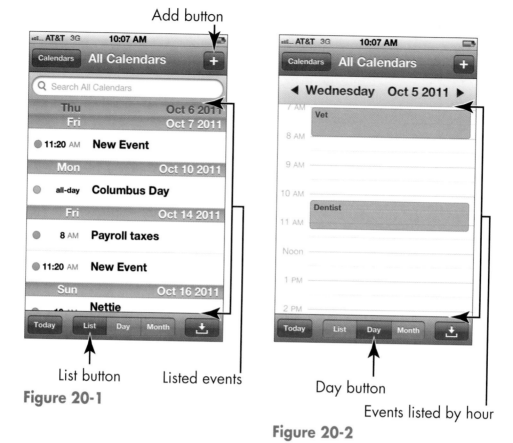

Add button

List button   Listed events

Figure 20-1

Day button

Events listed by hour

Figure 20-2

6. To jump back to today, tap the Today button in the bottom-left corner of Calendar.

To view any invitation that you accepted, which placed an event on your calendar, tap the Invitations button (it looks like a little inbox with an arrow pointing to it in the lower-right corner), and a list of invitations is displayed.

Selected day

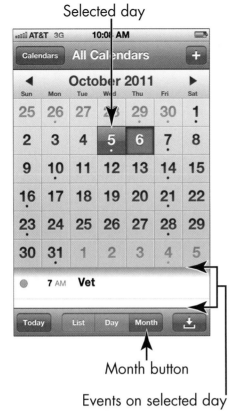

Month button

Events on selected day

Figure 20-3

Add Calendar Events

1. With any view displayed, tap the Add button (refer to **Figure 20-3**) to add an event. The Add Event dialog, shown in **Figure 20-4**, appears.

2. Enter a title for the event and, if you want, a location.

3. Tap the Starts/Ends field; the Start & End dialog, shown in **Figure 20-5**, is displayed.

4. Place your finger on the date, hour, minute, or AM/PM column and move your finger to scroll up or down. If you want to change the Time Zone, tap that field, begin to enter a new location, and then tap the location in the suggestions that appear. When each item is set correctly, tap Done. (Note that, if the event will last all day, you can simply tap the All-Day On/Off button and forget about setting start and end times.)

Figure 20-4

Figure 20-5

5. If you want to add notes, use your finger to scroll down in the Add Event dialog and tap in the Notes field. Type your note, and then tap the Done button to save the event.

 You can edit any event at any time by simply tapping it in any view of your calendar and, when the details are displayed, tap Edit. The Edit Event dialog appears, offering the same settings as the Add Event dialog. Tap the Done button to save your changes after you've made them.

Create Repeating Events

1. If you want an event to repeat, such as a weekly or monthly appointment, you can set a repeating event.

With any view displayed, tap the Add button to add an event. The Add Event dialog (refer to **Figure** 20-4) appears.

2. Enter a title and location for the event and set the start and end dates and times, as shown in the earlier task "Add Calendar Events."

3. Tap the Repeat field; the Repeat dialog, shown in **Figure 20-6,** is displayed.

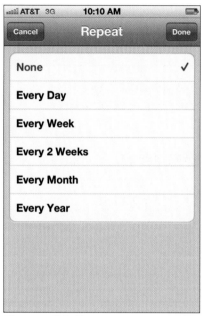

Figure 20-6

4. Tap a preset time interval: Every Day, Week, 2 Weeks, Month, or Year.

5. Tap Done. You return to the Add Event dialog.

6. Tap Done again to save your repeating event.

 Other calendar programs may give you more control over repeating events; for example, you might be able to make a setting to repeat an event every Tuesday. If you want a more robust calendar feature, you might consider setting up your appointments in iCal or Outlook and syncing them to iPhone. But, if you want to create a repeating event in iPhone's Calendar app, simply add the first event on a Tuesday and make it repeat every week. Easy, huh?

Add Alerts

1. If you want your iPhone to alert you when an event is coming up, you can use the Alert feature. First tap the Settings icon on the Home screen and choose Sounds.

2. Scroll down to Calendar Alerts and tap it; then tap any Alert Tone, which plays the tone for you. When you've chosen the alert tone you want, tap Sounds to return to Sounds settings.

Tap the Home button and then tap Calendar and create an event in your calendar or open an existing one for editing, as covered in earlier tasks in this chapter.

3. In the Add Event (refer to **Figure 20-4**) or Edit dialog, tap the Alert field. The Event Alert dialog appears, as shown in **Figure 20-7**.

4. Tap any preset interval, from 5 Minutes to 2 Days Before or At Time of Event. (Remember that you can scroll down in the dialog to see more options.)

5. Tap Done to save the alert, and note that the Alert setting is shown in the Edit dialog (see **Figure 20-8**).

6. Tap Done in the Edit dialog to save all settings.

7. Tap the Day button to display Day view of the date of your event; note that the alert and timeframe are listed under the event in that view.

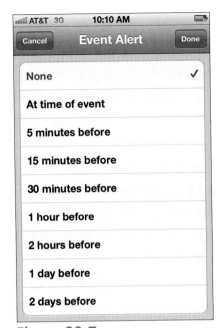

Figure 20-7

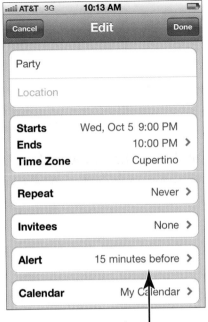

The alert setting for an event

Figure 20-8

 If you work for an organization that uses a Microsoft Exchange account, you can set up your iPhone to receive and respond to invitations from colleagues in your company. When somebody sends an invitation that you accept, it appears on your calendar. Check with your company network administrator (who will jump at the chance to get her hands on your iPhone) or the *iPhone User Guide* to set up this feature if it sounds useful to you.

Subscribe to and Share Calendars

1. If you use a calendar available from an online service such as Yahoo! or Google, you can subscribe to that calendar to read events saved there on your iPhone. Note that you can only read, not edit, these events. Tap the Settings icon on the Home screen to get started.

2. Tap the Mail, Contacts, Calendars option. The Mail, Contacts, Calendars settings pane appears.

3. Tap Add Account. The Add Account options, shown in **Figure 20-9**, appear.

4. Tap an e-mail choice, such as Gmail or Yahoo!.

5. In the dialog that appears (see **Figure 20-10**), enter your name, e-mail address, and e-mail account password.

Figure 20-9

Figure 20-10

6. Tap Save. iPhone verifies your address.

7. Your iPhone retrieves data from your calendar at the interval you have set to fetch data. To review these settings, tap the Fetch New Data option in the Mail, Contacts, Calendars dialog.

8. In the Fetch New Data dialog that appears (see **Figure 20-11**), be sure that the Push option's On/Off button reads *On* and then choose the option you prefer for how frequently data is pushed to your iPhone.

Make sure this is set to On

Figure 20-11

 If you use Microsoft Outlook's calendar or iCal on your main computer, you can sync it to your iPhone calendar to avoid having to reenter event information. To do this, use iCloud settings to sync automatically (see Chapter 3) or connect your iPhone to your computer with the Dock Connector to USB Cable and use settings in your iTunes account to sync with calendars — or use Wi-Fi syncing if you eschew cables. Click the Sync button, and your calendar settings will be shared between your computer and iPhone (in both directions). Read more in Chapter 3 about working with iTunes to manage your iPhone content.

 If you store birthdays for people in the Contacts app, the Calendar app then displays these when the day comes around so you won't forget to pass on your congratulations!

 Though you can have calendar events pushed to you or synced from multiple e-mail accounts, remember that having data pushed to your iPhone may drain your battery more quickly.

Delete an Event

1. When an upcoming luncheon or meeting is canceled, you should delete the appointment. With Calendar open, tap an event. Then tap the Edit button in the dialog that appears (see **Figure 20-12**). The Edit dialog opens.

2. In the Edit dialog, tap the Delete Event button at the bottom (see **Figure 20-13**).

Tap this button

Figure 20-12

Tap this button

Figure 20-13

3. If this is a repeating event, you have the option to delete this instance of the event or this and all future instances of the event (see **Figure 20-14**). Tap the button for the option you prefer. The event is deleted, and you return to Calendar view.

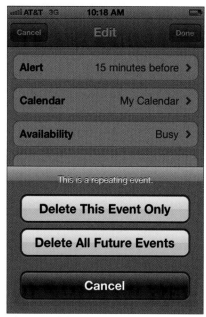

Figure 20-14

 If an event is moved but not canceled, you don't have to delete the old one and create a new one: Simply edit the event to change the day and time in the Event dialog.

Working with Reminders and Notifications

Chapter 21

With the arrival of iOS 5, the Reminders app and the Notification Center feature appeared, warming the hearts of those who need help remembering all the details of their lives.

Reminders is a kind of to-do list that lets you create tasks and set reminders so you don't forget them.

You can even be reminded to do things when you arrive at or leave a location. For example, you can set a reminder so that when your iPhone detects that you've left the location of your golf game, an alert reminds you to go buy milk, or when you arrive at your cabin, iPhone reminds you to turn on the water . . . you get the idea.

Notifications allows you to review all the things you should be aware of in one place, such as mail messages, text messages, calendar appointments, and alerts.

In this chapter, you discover how to set up and view tasks in Reminders and how the new Notification Center can centralize all your alerts in one easy-to-find place.

Create a Task in Reminders

1. Creating a task in Reminders is pretty darn simple. Tap Reminders on the Home screen.

2. On the screen that appears (see **Figure 21-1**), tap the Add button to add a task. The onscreen keyboard appears.

Tap here to create a new task

Figure 21-1

3. Enter a task name or description using the onscreen keyboard and tap the Return button. The new task is added to the Reminders list.

 You can't add details about a task when you enter it, only a descriptive name. To add details about timing and so forth, see the next task.

Edit Task Details

1. Tap a task to open the Details dialog. To see all the available options, tap Show More to display the choices shown in **Figure 21-2**. (Note that I deal with reminder settings in the following task.)

2. Tap Priority, choose None, Low, Medium, or High from the choices that appear, and then tap Done.

3. Tap Notes and enter any notes about the event (see **Figure 21-3**).

Enter notes here

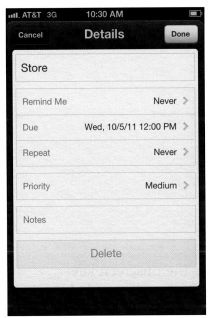

Figure 21-2

Figure 21-3

4. Tap Done to save the task.

With this version of the app, priority settings don't do much: They don't set a flag of any kind on a task in a list, nor do they reorder tasks to show priority tasks first. You can see the priority of a task only by displaying its details.

Schedule a Reminder by Time or Location

1. One of the major features of Reminders is to remind you of upcoming tasks. To set a reminder, tap a task.

2. In the dialog that appears (refer to **Figure 21-2**), tap Remind Me and, in the dialog that appears, tap the On/Off field for either the On a Day field (for a time-based reminder) or the At a Location field (for a location-based reminder) to turn it on.

3. Tap the date field to display the settings shown in **Figure 21-4.**

Date field

Figure 21-4

4. Tap and flick the day, hour, and minutes fields to scroll to the date and time for the reminder.

5. Tap Done to save the settings for the reminder.

 If you want a task to repeat with associated reminders, tap the Repeat field, and from the dialog that appears tap Every Day, Week, 2 Weeks, Month, or Year (for those annual meetings or great holiday get-togethers with the gang). Tap Done twice to save the setting.

Display Reminders as a List or by Date

1. You can display various lists of reminders (these include Reminders, tasks that come from My Calendar (the calendar associated with any e-mail account you've added to and synced with your iPhone), and Completed tasks, and lists you create yourself, as explained later in this chapter), or display reminders by date, which is a great help in planning your schedule. Tap Reminders and then tap the List button. A list of reminders appears, as shown in **Figure 21-5.**

2. Tap the Date tab and you see the tasks for the current date, as shown in **Figure 21-6.**

 You can scroll the monthly calendar display to show months in the past or future, and tap any date to show its tasks in the daily list on the right. Also with the Date view shown, tap a number in the list of dates along the bottom of the screen to go to that date.

Figure 21-5

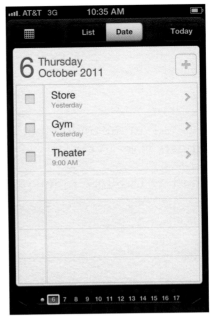

Figure 21-6

Create a List

1. You can create your own lists of tasks to help you keep different parts of your life organized. Tap Reminders on the Home screen to open it.

2. Tap the List View button (the button with the three lines on it, located to the left of the List button) to display the List view, and then tap the Edit button.

3. Tap Create New List, enter the name of the list using the onscreen keyboard, and then tap the Done button.

4. Tap Done to save the list and return to the List view. If you tap the List View button and then the name of the list you just created, a new blank sheet appears with that title. Tap the Add button to add new tasks to the list.

Sync with Other Devices and Calendars

1. To determine which tasks are brought over from other devices or calendars such as Outlook or iCal, tap the Settings button on the Home screen.

2. Tap iCloud. In the dialog that appears, be sure that Reminders is set to On.

3. Tap Settings to return to the main settings list, and then tap Mail, Contacts, Calendars and scroll down to the Reminders category.

4. Tap the Sync field and then choose how far back to sync Reminders (see **Figure 21-7**).

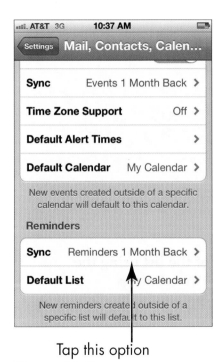

Tap this option

Figure 21-7

Note that to make all these settings work you should set up your default Calendar in the Calendar settings, and set up your iCloud account under Accounts in the Mail, Contacts, Calendar settings.

Mark as Complete or Delete a Reminder

1. You may want to mark a task as completed or just delete it entirely. With Reminders open and the list of tasks displayed, tap the check box to the left of a task to mark it as complete. When you tap the Completed category, the task now appears there and is removed from the Reminders category.

2. To delete a reminder, with the list of tasks displayed, tap a task to open the Details dialog shown in **Figure 21-8.**

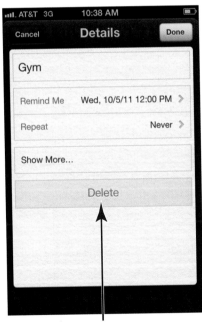

Tap this option

Figure 21-8

3. Tap Delete. In the confirming dialog, tap Delete again.

Set Notification Types

1. Notification Center is on by default, but you don't have to include every type of notification there if you don't want to; for example you might never want to be notified of incoming messages but always want to have reminders listed here — it's up to you. There are some settings you can make to control what types of notifications are included in it. Tap Settings, and then tap Notifications.

2. In the settings that appear (see **Figure 21-9**), note that there is a list of items included in Notification Center and a list of items not included. For example, Messages and Reminders may be included, but alerts in game apps may not.

3. Tap any item and, in the settings that appear (see **Figure 21-10**), tap the On/Off button to have that item included or excluded from Notification Center.

Include or exclude from Notification Center

Figure 21-9

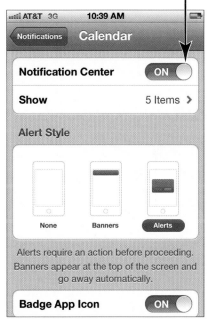

Figure 21-10

4. Tap an Alert Style to have no alert, a banner across the top of the screen, or a boxed alert appear. If you choose Banner, it will appear and then disappear automatically. If you choose Alert, you have to take an action to dismiss the alert when it appears.

5. Badge App Icon is a feature that places a red circle and number on icons on your Home screens representing alerts associated with those apps. To turn this feature off, tap the On/Off button for Badge App Icon.

6. If you want to be able to view alerts when the Lock Screen is displayed, turn on the View in Lock Screen setting. When you've finished making settings for an individual app, tap the Settings button to go back to the Notifications settings.

View Notification Center

1. Once you've made settings for what should appear in Notification Center, you'll regularly want to take a look at those alerts and reminders. From any screen, tap the black status bar on top and drag down to display Notification Center (see **Figure 21-11**).

2. Note that items are divided into lists by type — for example, you'll see items categorized as Reminders, Mail, Calendar, and so on.

3. To close Notification Center, tap the three lines in the bottom center of the screen and drag up toward the status bar.

 To determine what is displayed in Notification Center, see the previous task.

Figure 21-11

You can display Stock and Weather Widgets in Notification Center by tapping Settings, Notifications, and making sure these appear in the "In Notification Center" list; if they don't, swipe down to the "Not In Notification Center" list, tap on the widget you want to display, and tap the "Notification Center" On/Off button.

Go to an App from Notification Center

1. You can easily jump from Notification Center to any app that caused an alert or reminder to be displayed. Tap the status bar and drag down to display Notification Center.

2. Tap any item; it opens in its originating app. If you've tapped a message such as an e-mail, you can then reply to the message using the procedure described in Chapter 11.

Clear Notifications

1. To get rid of old notifications for an app, tap the status bar and drag down to display Notification Center.

2. Tap the pale gray X to the right of a category of notifications, such as Mail. The button changes to read Clear.

3. Tap the Clear button, and the notification is removed from Notification Center.

Making Notes

*N*otes is the included app that you can use to do everything from jotting down notes at meetings to keeping to-do lists. It isn't a robust word processor (such as Apple Pages or Microsoft Word) by any means, but for taking notes on the fly, jotting down shopping lists, or writing a few pages of your novel-in-progress while you sit and sip a cup of tea on your deck, it's a great option.

In this chapter, you see how to enter and edit text in Notes and how to manage those notes by navigating among them, searching for content, or e-mailing or deleting them.

Get ready to . . .

Open a Blank Note

1. To get started with Notes, tap the Notes app icon on the Home screen. If you've never used Notes, it opens with a new, blank note displayed. (If you have used Notes, it opens to the last note you were working on. If that's the case, you might want to jump to the next task to display a new, blank note.) You see the view shown in **Figure 22-1**.

2. Tap the blank page, or tap the New Note button (refer to **Figure 22-1**). The onscreen keyboard, shown in **Figure 22-2**, appears.

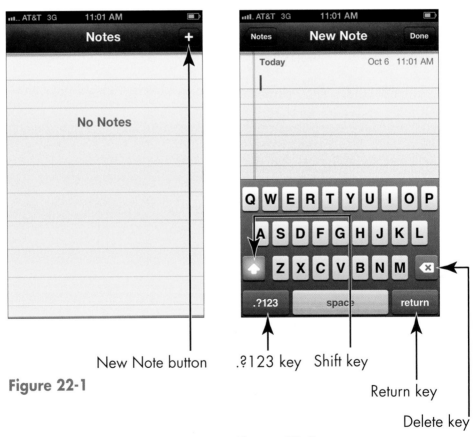

New Note button

Figure 22-1

.?123 key   Shift key

Return key

Delete key

Figure 22-2

3. Tap keys on the keyboard to enter text. If you want to enter numbers or symbols, tap the key labeled *.?123* on the keyboard (refer to **Figure 22-2**). The numerical keyboard, shown in **Figure 22-3,** appears. Whenever you want to return to the alphabetic keyboard, tap the key labeled *ABC*.

Done button

The ABC key

Figure 22-3

4. To capitalize a letter, tap the Shift key (refer to **Figure 22-2**) and then tap the letter. You can enable the Enable Caps Lock feature in the General Keyboard Settings, or you can turn caps lock on by double-tapping the Shift key; tap the Shift key once to turn the feature off.

5. When you want to start a new paragraph or a new item in a list, tap the Return key (refer to **Figure 22-2**).

6. To edit text, tap the text you want to edit and either use the Delete key (refer to **Figure 22-2**) to delete text to the left of the cursor or type new text.

When you have the numerical keyboard displayed (refer to **Figure 22-3**), you can tap the key labeled #+= to access more symbols, such as the percentage sign or the euro symbol, or additional bracket styles.

No need to save a note — it's kept automatically until you delete it.

Create a New Note

1. With one note open, to create a new note, tap the Done button (refer to **Figure 22-3**) and then tap the New Note button — the one with the plus sign (+) on it — in the top-right corner.

2. A new, blank note appears (refer to **Figure 22-1**). Enter and edit text as described in the previous task.

You can tap the Notes button to see a list of all notes you've created. Tap an individual note to display it again.

Use Copy and Paste

1. The Notes app includes two essential editing tools you're probably familiar with from other word processors: copy and paste. With a note displayed, press and hold your finger on a word. The toolbar shown in **Figure 22-4** appears.

2. Tap the Select button. The toolbar shown in **Figure 22-5** appears.

Tap this button

Tap this button

Figure 22-4

Figure 22-5

3. Tap the Copy button.

4. Press and hold your finger in the document at the spot where you want to place the copied text.

5. On the toolbar that appears (see **Figure** 22-6), tap the Paste button. The copied text appears.

 If you want to select all text in a note to either delete or copy it, tap the Select All button on the toolbar shown in **Figure 22-4.** The Cut/Copy/Paste/Suggest toolbar appears, which you can use to deal with the selected text. To extend a selection to adjacent words, press one of the little handles that extends from the selection and drag to the left or right. To get an alternate spelling suggestion, you can tap Suggest.

Tap this button

Figure 22-6

 To delete text, you can use the Select or Select All command and then press the Delete key on the onscreen keyboard.

Display the Notes List

1. Tap the Notes app icon on the Home screen to open Notes.

2. Tap the Notes button in the top-left corner of the screen; the notes list appears, as shown in **Figure 22-7**. This list is organized chronologically (the date isn't indicated if you created the note today).

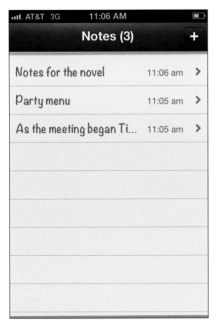

₋₋₋ᴵᴵ. AT&T 3G	11:06 AM	🔳
	Notes (3)	**+**
Notes for the novel	11:06 am	>
Party menu	11:05 am	>
As the meeting began Ti...	11:05 am	>

Figure 22-7

3. Tap any note on the list to display it.

 Notes names your note, using the first line of text. If you want to rename a note, first display the note, tap at the end of the first line of text, and then tap the Delete key on your onscreen keyboard to delete the old title. Enter a new title; it's reflected as the name of your note in the notes list.

Move among Notes

1. You have a couple of ways to move among notes you've created. Tap the Notes app icon on the Home screen to open Notes.

2. With the notes list displayed (see the previous task), tap a note to open it.

3. To move among notes, tap the Next or Previous button (the right- or left-facing arrow at the bottom of the Notes pad, as shown in **Figure 22-8**).

Tap either of these buttons

Figure 22-8

 Because Notes lets you enter multiple notes with the same title — which can cause confusion — name your notes uniquely!

Search for a Note

1. You can search to locate a note that contains certain text. The Search feature lists only notes that contain your search criteria; it doesn't highlight and show you every instance of the word or words you enter. Tap the Notes app icon on the Home screen to open Notes.

2. Tap the Notes button in the top-left corner to display the notes list if it isn't already displayed.

3. Press your finger on the middle of the screen and swipe down. The Search field appears above the notes list, as shown in **Figure 22-9**.

4. Tap in the Search field. The onscreen keyboard appears.

5. Begin to enter the search term. All notes that contain matching words appear on the list, as shown in **Figure 22-10**.

The Search field

Notes containing the search term

Figure 22-9   **Figure 22-10**

6. Tap a note to display it and then locate the instance of the matching word the old-fashioned way — by skimming to find it.

E-mail a Note

1. If you want to share what you wrote with a friend or colleague, you can easily e-mail the contents of a note. With a note displayed, tap the Menu button at the bottom of the screen, as shown in **Figure 22-11**.

2. Tap Email. In the e-mail form that appears (see **Figure 22-12**), type one or more e-mail addresses in the appropriate fields. At least one e-mail address must appear in the To field.

Tap this button

Figure 22-11

Enter e-mail addresses here

Figure 22-12

3. If you need to make changes to the subject or message, tap in either area and make the changes.

4. Tap the Send button, and your e-mail is on its way.

If you want to print a note, in Step 2, choose Print rather than Email. Complete the Printer Options dialog by designating an AirPrint-enabled wireless printer (or shared printer on a network that you can access using AirPrint) and how many copies to print, and then tap Print.

 You can tap the button with a plus sign (+) on it in the top-right corner of the e-mail message form to display your contacts list and choose recipients from it. This method works only with contacts for whom you have added an e-mail address. See Chapter 5 for more about using the Contacts app.

 To cancel an e-mail message and return to Notes without sending it, tap the Cancel button in the e-mail form and then tap Don't Save on the menu that appears. To leave a message but save a draft so that you can finish and send it later, tap Cancel and then tap Save. The next time you tap the e-mail button with the same note displayed in Notes, your draft appears.

Delete a Note

1. There's no sense in letting your notes list get cluttered, making it harder to find the ones you need. When you're done with a note, it's time to delete it. Tap the Notes app icon on the Home screen to open Notes.

2. Tap a note in the notes list to open it.

3. Tap the Trash Can button, shown in **Figure 22-13**.

4. Tap the Delete Note button that appears (see **Figure 22-14**). The note is deleted.

Tap this button

Figure 22-13

Tap this button

Figure 22-14

Notes is a nice little application, but it's limited. It offers no formatting tools or ways to print the content you enter. You can't paste pictures into Notes. (You can try, but it won't work: Only the filename appears, not the image.) So, if you've made some notes and want to graduate to building a more robust document in a word processor, you have a couple of options. One way is to buy the Pages word processor application for iPad, which costs about $9.99, and copy your note (using the copy-and-paste feature discussed earlier in this chapter). Alternatively, you can send the note to yourself in an e-mail message or sync it to your computer. Open the e-mail or note and copy and paste its text into a full-fledged word processor, and you're good to go.

Troubleshooting and Maintaining Your iPhone

*i*Phones don't grow on trees — they cost a pretty penny. That's why you should learn how to take care of your iPhone and trouble-shoot any problems it might have so that you get the most out of it.

In this chapter, I provide some advice about the care and maintenance of your iPhone, as well as tips about how to solve common prob-lems, update iPhone system software, and even reset the iPhone if something goes seriously wrong. In case you lose your iPhone, I even tell you about a new feature that helps you find it — or even disable it if it has fallen into the wrong hands. Finally, you get information about backing up your iPhone settings and content using iCloud.

Get ready to . . .

Keep the iPhone Screen Clean

If you've been playing with your iPhone, you know (despite Apple's claim that the iPhone has a fingerprint-resistant screen) that it's a fingerprint magnet. Here are some tips for cleaning your iPhone screen:

➡ **Use a dry, soft cloth.** You can get most fingerprints off with a dry, soft cloth such as the one you use to clean your eyeglasses or a cleaning tissue that's lint- and chemical-free. Or try products used to clean lenses in labs such as Kimwipes or Kaydry, which you can get from several major retailers such as Amazon.

➡ **Use a slightly dampened soft cloth.** To get the surface even cleaner, slightly dampen the soft cloth. Again, make sure that whatever cloth material you use is free of lint.

➡ **Remove the cables.** Turn off your iPhone and unplug any cables from it before cleaning the screen with a moistened cloth.

➡ **Avoid too much moisture.** Avoid getting too much moisture around the edges of the screen, where it can seep into the unit.

➡ **Never use household cleaners.** They can degrade the coating that keeps the iPhone screen from absorbing oil from your fingers.

 Do *not* use premoistened lens-cleaning tissues to clean your iPhone screen. Most brands of wipe contain alcohol, which can damage the screen's coating.

Protect Your Gadget with a Case

Your screen isn't the only element on the iPhone that can be damaged, so consider getting a case for it so you can carry it around the house or around town safely. Besides providing a bit of padding if you drop the device, a case makes the iPhone less slippery in your hands, offering a better grip when working with it.

Several types of cases are available. You can choose covers from manufacturers such as Griffin (www.griffintechnology.com) that come in materials ranging from leather to silicone.

Cases range from a few dollars to $70 or more for leather (with some outrageously expensive designer cases upward of $500). Some provide a cover for the screen and back, and others protect only the back and sides. If you carry your iPhone around much, consider a case with a screen cover to provide better protection for the screen or use a screen overlay, such as InvisibleShield from Zagg (www.zagg.com).

Extend Your iPhone's Battery Life

The much-touted battery life of the iPhone is a wonderful feature, but you can do some things to extend it even further. Here are a few tips to consider:

⟹ **Keep tabs on remaining battery life.** You can estimate the amount of remaining battery life by looking at the Battery icon on the far-right end of the Status bar, at the top of your screen.

⟹ **Use standard accessories to charge your iPhone most effectively.** When connected to a modern Mac or Windows computer for charging, the iPhone can slowly charge; however, the most effective way to charge your iPhone is to plug it into the wall outlet using the Dock Connector to USB Cable and the 10W USB Power Adapter that come with your iPhone.

⟹ The fastest way to charge the iPhone is to turn it off while charging it.

⟹ The Battery icon on the Status bar lets you know when the charging is complete.

Your iPhone battery is sealed in the unit, so you can't replace it, as you can with many laptops or cellphone batteries. If the battery is out of warranty, you have to fork over about $79 to get a new one. See the "Get Support" task, later in this chapter, to find out where to get a replacement battery.

Find Out What to Do with a Nonresponsive iPhone

If your iPhone goes dead on you, it's most likely a power issue, so the first thing to do is to plug the Dock Connector to USB Cable into the 10W USB Power Adapter, plug the 10W USB Power Adapter into a wall outlet, plug the other end of the Dock Connector to USB Cable into your iPhone, and charge the battery.

Another thing to try — if you believe that an app is hanging up the iPhone — is to press the Sleep/Wake button for a couple of seconds. Then press and hold the Home button. The app you were using should close.

You can always try the old reboot procedure, too: On the iPhone, you press the Sleep/Wake button on top until the red slider appears. Drag the slider to the right to turn off your iPhone. After a few moments, press the Sleep/Wake button to boot up the little guy again.

If the situation seems drastic and none of these ideas works, try to reset your iPhone. To do this, press the Sleep/Wake button and the Home button at the same time until the Apple logo appears onscreen.

Update Software

1. Apple occasionally updates the iPhone system software to fix problems or offer enhanced features. You should occasionally check for an updated version (say, every month). Start by connecting your iPhone to your computer.

2. On your computer, open the iTunes software you installed. (See Chapter 3 for more about this topic.)

3. Click on your iPhone in the iTunes source list on the left.

4. Click the Summary tab, shown in **Figure 23-1**.

Click on your iPhone...

then click the Summary tab

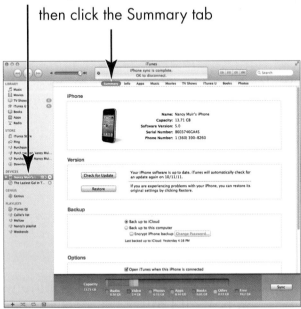

Figure 23-1

5. Click the Check for Update button. iTunes displays a message telling you whether a new update is available.

6. Click the Update button to install the newest version.

Note that with iOS 5 and iCloud you can also tap Settings, General, and then tap Software Update. Your phone will go out and look for and update and install it for you.

If you're having problems with your iPhone, you can use the Update feature to try to restore the current version of the software. Follow the preceding set of steps, and then click the Restore button instead of the Update button in Step 6.

If you've chosen to back up and restore iPhone via iCloud when you first set up the device, restoring and updating your device happens automatically.

Restore the Sound

On the morning I wrote this chapter, as my husband puttered with our iPhone, its sound suddenly (and ironically) stopped working. We gave ourselves a quick course in sound recovery, so now I can share some tips with you. Make sure that

➡ **You haven't touched the volume control buttons on the side of your iPhone.** They're on the left side of the phone. Be sure not to touch the volume decrease button and inadvertently lower the sound to a point where you can't hear it.

➡ **You haven't flipped the Silent switch.** Moving the switch located on the left side above the volume buttons mutes sound on the iPhone.

➡ **The speaker isn't covered up.** It may be covered in a way that muffles the sound.

➡ **A headset isn't plugged in.** Sound doesn't play over the speaker and the headset at the same time.

➡ **The volume limit is set to Off.** You can set up the volume limit in the Music settings to control how loudly your music can play (which is useful if you have teenagers around). Tap the Settings icon on the Home screen and then, on the screen that displays, tap Music and use the Volume Limit control (see **Figure 23-2**) to turn off the volume limit.

When all else fails, reboot. This strategy worked for us — just press the Sleep/Wake button until the red slider appears. Press and drag the slider to the right. After the iPhone turns off, press the Sleep/Wake button again until the Apple logo appears, and you may find yourself back in business, sound-wise.

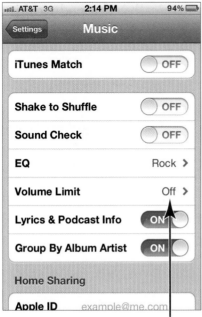

Make sure this is set to Off

Figure 23-2

Get Support

Apple is known for its helpful customer support, so if you're stuck, I definitely recommend that you try it out. Here are a few options you can explore for getting help:

➡ **The Apple Store:** Go to your local Apple Store (if one is handy) to see what the folks there might know about your problem.

➠ **The Apple support website:** It's at `www.apple.com/support/iphone`. You can find online manuals, discussion forums, and downloads, and you can use the Apple Expert feature to contact a live support person by phone.

➠ **The *iPhone User Guide:*** You can use the bookmarked manual on the Safari browser to view the user guide that comes with your iPhone.

➠ **The Apple battery replacement service:** If you need repair or service for your battery, visit `www.apple.com/batteries/replacements.html`. Note that your warranty provides free battery replacement if the battery level dips below 50 percent and won't go any higher during the first year you own it. If you purchase the AppleCare service agreement, this is extended to two years.

 Apple recommends that you have your iPhone battery replaced only by an Apple Authorized Service Provider.

Find a Missing iPhone

You can take advantage of the Find My iPhone feature to pinpoint the location of your iPhone. This feature is extremely handy if you forget where you left your iPhone or someone steals it. Find My iPhone not only lets you track down the critter but also lets you wipe out the data contained in it if you have no way to get the iPhone back.

Follow these steps to set up the Find My iPhone feature:

1. Tap the Settings icon on the Home screen.

2. In the Settings pane, tap iCloud.

3. In the iCloud settings, tap the On/Off button for Find My iPhone to turn the feature on (see **Figure** 23-3).

Set this option to On

Figure 23-3

4. From now on, if your iPhone is lost or stolen, you can go to http://iCloud.com from your computer and enter your ID and password.

5. The Find My iPhone screen appears with your iPhone's location noted on a map.

6. To wipe information from the iPhone, click the Remote Wipe button. To lock the iPhone from access by others, click the Remote Lock button.

Remote Wipe will delete all data from your iPhone, including contact information and content such as music.

You can also click Display a Message or Play a Sound to send whoever has your iPhone a note saying how to return it to you — or that the police are on their way, if it has been stolen! If you choose to play a sound, it plays for two minutes, helping you track down your phone if you left it on top of the refrigerator or anybody holding your iPhone who is within earshot.

Back up to iCloud

You used to be able to back up your iPhone content using only iTunes, but with Apple's introduction of iCloud, you can back up via a Wi-Fi network to your iCloud storage. You get 5GB of storage (not including iTunes-bought music, iTunes Match and Photo Stream content, video, apps, and e-books) for free or you can pay for increased storage (10GB for $20 per year, 20GB for $40 per year, or 50GB for $100 per year).

1. To perform a backup to iCloud, first set up an iCloud account (see Chapter 3 for details on creating an iCloud account) and then tap Settings on the Home Screen.

2. Tap iCloud and then tap Storage & Backup (see **Figure 23-4**).

3. In the pane that appears (see **Figure 23-5**) tap the iCloud Backup On/Off switch to enable automatic backups. To perform a manual backup, tap Back Up Now. A progress bar shows how your backup is moving along.

Tap this option

Figure 23-4

Set this to On

Figure 23-5

Index